SpringerBriefs in Statistics

For further volumes:
http://www.springer.com/series/8921

Debasis Kundu · Swagata Nandi

Statistical Signal Processing

Frequency Estimation

 Springer

Debasis Kundu
Department of Mathematics and Statistics
Indian Institute of Technology
Kanpur, Uttar Pradesh 208016
India

Swagata Nandi
Theoretical Statistics and Mathematics Unit
Indian Statistical Institute
7, S.J.S. Sansanwal Marg
New Delhi 110016
India

ISSN 2191-544X   ISSN 2191-5458   (electronic)
ISBN 978-81-322-0627-9 ISBN 978-81-322-0628-6   (eBook)
DOI 10.1007/978-81-322-0628-6
Springer New Delhi Heidelberg New York Dordrecht London

Library of Congress Control Number: 2012938023

Printed on acid-free paper

Springer is part of Springer Science+Business Media (www.springer.com)

To my parents
                D. Kundu

To my parents
                S. Nandi

Preface

We have worked on our Ph.D. theses on *Statistical Signal Processing* although in a gap of almost 15 years. The first author was introduced to the area by his Ph.D. supervisor Professor C. R. Rao, while the second author was introduced to this topic by the first author. It has been observed that frequency estimation plays an important role in dealing with different problems in the area of *Statistical Signal Processing*, and both the authors have spent a significant amount of their research career dealing with this problem for different models associated with Statistical Signal Processing.

Although an extensive amount of literature is available in the engineering literature dealing with the frequency estimation problem, not much attention has been paid to the statistical literature. The book by Quinn and Hannan [1] is the only book dealing with the problem of frequency estimation written for the statistical community. We were thinking of writing a review article on this topic for quite sometime. In this respect, the invitation from Springer to write a Springer Brief on this topic came as a pleasant surprise to us.

In this Springer Brief, we provide a review of the different methods available till date dealing with the problem of frequency estimation. We have not attempted an exhaustive survey of frequency estimation techniques. We believe that would require separate books on several topics themselves. Naturally, the choice of topics and examples are based, in favor of our own research interests. The list of references is also far from complete.

We have kept the mathematical level quite modest. Chapter 4 mainly deals with somewhat more demanding asymptotic theories, and this chapter can be avoided during the first reading without losing any continuity. Senior undergraduate level mathematics should be sufficient to understand the rest of the chapters. Our basic goal to write this Springer Brief is to introduce the challenges of the frequency estimation problem to the statistical community, which are present in different areas of science and technology. We believe that statisticians can play a major role in solving several problems associated with frequency estimation. In Chap. 8, we have provided several related models, where there are several open issues which need to be answered by the scientific community.

Every book is written with a specific audience in mind. This book definitely cannot be called a textbook. It has been written mainly for senior undergraduate and graduate students specializing in Mathematics or Statistics. We hope that this book will motivate students to pursue higher studies in the area of Statistical Signal Processing. This book will be helpful to young researchers who want to start their research career in the area of Statistical Signal Processing. We will consider our efforts to be worthy if the target audience finds this volume useful.

Kanpur, January 2012   Debasis Kundu
Delhi, January 2012   Swagata Nandi

Reference

1. Quinn, B.G., & Hannan, E.J. (2001). *The estimation and tracking of frequency.* New York: Cambridge University Press.

Acknowledgments

The authors would like to thank their respective Institutes and to all who contributed directly and indirectly to the production of this monograph. The first author would like to thank his wife Ranjana for her continued support and encouragement.

Contents

Abbreviations

1-D	One-dimensional
2-D	Two-dimensional
3-D	Three-dimensional
AIC	Akaike's information criterion
ALSE(s)	Approximate least squares estimator(s)
AM	Amplitude modulated
AR	Autoregressive
ARIMA	Autoregressive integrated moving average
ARMA	Autoregressive moving average
BIC	Bayesian information criterion
CV	Cross-validation
ECG	Electrocardiograph
EDC	Efficient detection criterion
EM	Expectation maximization
ESPRIT	Estimation of signal parameters via rotational invariance technique
EVLP	Equivariance linear prediction method
i.i.d.	Independent and identically distributed
Im(z)	Imaginary part of a complex number z
ITC	Information theoretic criterion
IQML	Iterative quadratic maximum likelihood
KIC	Kundu's information criterion
LSE(s)	Least squares estimator(s)
MA	Moving average
MDL	Minimum description length
MFBLP	Modified forward backward linear prediction method
MLE(s)	Maximum likelihood estimator(s)
NSD	Noise space decomposition
PE(s)	Periodogram estimator(s)
QIC	Quinn's information criterion
QT	Quinn and Thomson

Re(z)	Real part of a complex number z
RGB	Red-green-blue
SIC	Sakai's information criterion
TLS-ESPRIT	Total least squares estimation of signal parameters via rotational invariance technique
WIC	Wang's information criterion
WLSE(s)	Weighted least squares estimator(s)

Symbols

$\arg(z)$	$\tan^{-1}\theta$ where $z = r\,e^{i\theta}$
$\mathcal{N}(a,\,b)^2$	Univariate normal distribution with mean a and variance b^2
$\mathcal{N}_p(\mathbf{0},\,\Sigma)$	p-variate normal distribution with mean vector $\mathbf{0}$ and dispersion matrix Σ
$X_n \xrightarrow{d} X$	X_n converges to X in distribution
$X_n \xrightarrow{p} X$	X_n converges to X in probability
a.s.	Almost surely
i.o.	Infinitely often
a.e.	Almost everywhere
$X_n \xrightarrow{a.s.} X$	X_n converges to X almost surely
$o(a_n)$	$(1/a_n)o(a_n) \to 0$ as $n \to \infty$
$O(a_n)$	$(1/a_n)O(a_n)$ is bounded as $n \to \infty$
$X_n = o_p(a_n)$	$\lim_{n\to\infty} P(|X_n/a_n| \geq \epsilon) - 0$, for every positive ϵ
$X_n = O_p(a_n)$	X_n/a_n is bounded in probability as $n \to \infty$
$\|\mathbf{x}\|$	$\sqrt{x_1^2 + \ldots + x_n^2}$, $\mathbf{x} = (x_1, x_2, \ldots, x_n)$
$\mathbb{R}$	Set of real numbers
$\mathbb{R}$	d-Dimensional Euclidean space
$\mathbb{C}$	Set of complex numbers
$|\mathbf{A}|$	Determinant of matrix $\mathbf{A}$
$X \overset{d}{=} Y$	X and Y are identically distributed

Chapter 1
Introduction

Signal processing may broadly be considered to involve the recovery of information from physical observations. The received signal is usually disturbed by thermal, electrical, atmospheric, or intentional interferences. Due to random nature of the signal, statistical techniques play important roles in analyzing the signal. Statistics is also used in the formulation of appropriate models to describe the behavior of the system, the development of an appropriate technique for the estimation of model parameters, and the assessment of the model performances. Statistical signal processing basically refers to the analysis of random signals using appropriate statistical techniques.

The main aim of this monograph is to introduce different signal processing models which have been used in analyzing periodic data, and different statistical and computational issues associated with them. We observe periodic phenomena everyday in our lives. The daily temperature of Delhi or the number of tourists visiting the famous Taj Mahal everyday or the ECG signal of a normal human being, clearly follow periodic nature. Sometimes, the observations/signals may not be exactly periodic on account of different reasons, but they may be nearly periodic. It should be clear from the following examples, where the observations are obtained from different disciplines, that they are nearly periodic. In Fig. 1.1, we provide the ECG signal of a healthy person. In Fig. 1.2, we present an astronomical data set which represents the daily brightness of a variable star on 600 successive midnights. Figure 1.3 represents a classical data set of the monthly international airline passengers from January 1953 to December 1960 and is collected from the Time Series Data Library http://www.robhyndman.info/TDSL.

The simplest periodic function is the sinusoidal function and it can be written in the following form:

$$y(t) = A \cos(\omega t) + B \sin(\omega t), \tag{1.1}$$

where $A^2 + B^2$ is known as the amplitude of $y(t)$ and ω is the frequency. In general, a *smooth* mean zero periodic function $y(t)$ can always be written in the form

D. Kundu and S. Nandi, *Statistical Signal Processing*, SpringerBriefs in Statistics, DOI: 10.1007/978-81-322-0628-6_1, © The Author(s) 2012

Fig. 1.1 ECG signal of a
healthy person

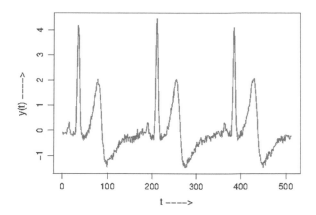

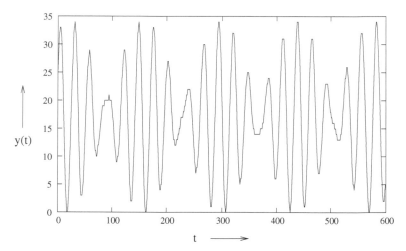

Fig. 1.2 Brightness of a variable star

$$y(t) = \sum_{k=1}^{\infty} \left[A_k \cos(k\omega t) + B_k \sin(k\omega t) \right], \tag{1.2}$$

and it is well known as the Fourier expansion of $y(t)$. From $y(t)$, A_k and B_k for
$k = 1, \ldots, \infty$, can be obtained uniquely. Unfortunately in practice, we hardly
observe *smooth* $y(t)$. Most of the times $y(t)$ is corrupted with noise as observed
in the above three examples; hence, it is quite natural to use the following model

$$y(t) = \sum_{k=1}^{\infty} \left[A_k \cos(k\omega t) + B_k \sin(k\omega t) \right] + X(t), \tag{1.3}$$

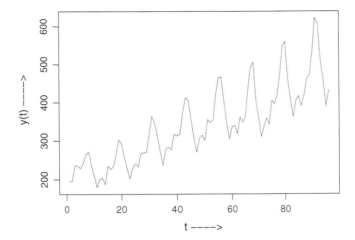

Fig. 1.3 Monthly international airline passengers during the period from January 1953 to December 1960

to analyze the noisy periodic signal. Here $X(t)$ is the noise component. It is impossible to estimate the infinite number of parameters in practice, hence (1.3) is approximated by

$$y(t) = \sum_{k=1}^{p} [A_k \cos(\omega_k t) + B_k \sin(\omega_k t)] + X(t), \qquad (1.4)$$

for some $p < \infty$. Due to this reason, quite often, the main problem boils down to estimating p, and A_k, B_k, ω_k for $k = 1, \ldots, p$, from the observed signal $\{y(t); t = 1, \ldots, n\}$.

The problem of estimating the parameters of model (1.4) from the data $\{y(t); t = 1, \ldots, n\}$ becomes a classical problem. Starting with the work of Fisher [1], this problem has received considerable attention because of its widescale applicability. Brillinger [2] discussed some of the very important real-life applications from different areas and provided solutions using the sum of sinusoidal model. Interestingly, but not surprisingly, this model has been used quite extensively in the signal processing literature. Kay and Marple [3] wrote an excellent expository article from the signal processor's point of view. More than 300 list of references can be found in Stoica [4] on this particular problem till that time. See also two other review articles by Prasad et al. [5] and Kundu [6] in this area. The monograph of Quinn and Hannan [7] is another important contribution in this area.

This problem has several different important and interesting aspects. Although, model (1.4) can be viewed as a non-linear regression model, this model does not satisfy the standard assumptions needed for different estimators to behave *nicely*. Therefore, deriving the properties of the estimators is an important problem. The usual consistency and asymptotic normality results do not follow from the general results. Moreover, finding the estimators of the unknown parameters is well known to be a numerically difficult problem. The problem becomes more complex if $p \geq 2$. Because of these reasons, this problem becomes challenging both from the theoretical and computational points of view.

Model (1.4) is a regression model, hence in the presence of independent and identically distributed (i.i.d.) error $\{X(t)\}$, the least-squares method seems to be a reasonable choice for estimating the unknown parameters. But interestingly, instead of the least-squares estimators (LSEs), traditionally the periodogram estimators (PEs) became more popular. The PEs of the frequencies can be obtained by finding the local maximums of the periodogram function $I(\omega)$, where

$$I(\omega) = \frac{1}{n} \left| \sum_{t=1}^{n} y(t) e^{-i\omega t} \right|^2. \tag{1.5}$$

Hannan [8] and Walker [9] independently first obtained the theoretical properties of the PEs. It is observed that the rate of convergence of the PEs of the frequencies is $O_p(n^{-3/2})$. Kundu [10] observed that the rates of convergence of the LSEs of the frequencies and amplitudes are $O_p(n^{-3/2})$ and $O_p(n^{-1/2})$, respectively. This unusual rate of convergence of the frequencies makes the model interesting from the theoretical point of view.

Finding the LSEs or the PEs is a computationally challenging problem. The problem is difficult because the least-squares surface as well as the periodogram surface of the frequencies are highly non-linear. There are several local optimums in both the surfaces. Thus, very good (close enough to the true values) initial estimates are needed for any iterative process to work properly. It is also well known that the standard methods like Newton–Raphson or Gauss–Newton do not work well for this problem. One of the common methods to find the initial guesses of the frequencies is to find the local maximums of the periodogram function $I(\omega)$, at the Fourier frequencies, that is, restricting the search space only at the discrete points $\{\omega_j = 2\pi j/n; j = 0, \ldots, n-1\}$. Asymptotically, the periodogram function has local maximums at the true frequencies. But unfortunately, if two frequencies are very close to each other, then this method may not work properly.

Just to see the complexity of the problem, consider the periodogram function of the following synthesized signal;

$$y(t) = 3.0 \cos(0.20\pi t) + 3.0 \sin(0.20\pi t)$$
$$+ 0.25 \cos(0.19\pi t) + 0.25 \sin(0.19\pi t) + X(t); \quad t = 1, \ldots, 75. \tag{1.6}$$

Here, $\{X(t); t = 1, \ldots, 75\}$ are i.i.d. normal random variables with mean 0 and variance 0.5. The periodogram function of the observed signal from model (1.6) is provided in Fig. 1.4. In this case clearly, two frequencies are not resolvable. It is not immediate how to choose initial estimates in this case to start any iterative process to find the LSEs or the PEs.

Because of this, several other techniques are available in practice, which attempt to find efficient estimators without using any iterative procedure. Most of the methods

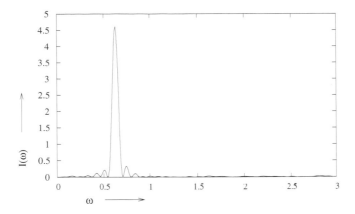

Fig. 1.4 Periodogram function of the synthesized data obtained from model (1.6)

make use of the recurrence relation formulation (1.7) obtained from the celebrated Prony's equation (see in Chap. 2);

$$\sum_{j=0}^{2p} c(j)y(t-j) = \sum_{j=0}^{2p} c(j)X(t-j); \quad t = 2p+1, \ldots, n, \qquad (1.7)$$

where $c(0) = c(2p) = 1, c(j) = c(2p - j)$ for $j = 0, \ldots, 2p$. Relation (1.7) is formally equivalent to saying that a linear combination of p sinusoidal signals can be modeled as an ARMA$(2p, 2p)$ process. The coefficients $c(1), \ldots, c(2p - 1)$ depend only on the frequencies $\omega_1, \ldots, \omega_p$ and they can be obtained uniquely from the $c(1), \ldots, c(2p - 1)$. Due to this relation, a variety of procedures have been evolved since 1970s on estimating the coefficients $c(j)$ for $j = 1, \ldots, 2p - 1$ from the observed signal $\{y(t); t - 1, \ldots, n\}$. From the estimated $c(1), \ldots, c(2p - 1)$, the estimates of $\omega_1, \ldots, \omega_p$ can be easily obtained. Since all these methods are non-iterative in nature, they do not demand any initial guesses. But, the frequency estimates produced by these methods are mostly $O_p(n^{-1/2})$, not $O_p(n^{-3/2})$. Therefore, their efficiency is much lower than that of LSEs or the PEs.

Another important problem is to estimate p, when it is unknown. Fisher [1] first treated this as a testing of hypothesis problem. Later, several authors attempted different information theoretic criteria, namely AIC, BIC, EDC, etc., or their variants. But choosing the proper penalty function seems to be a really difficult problem. Cross validation technique has also been used to estimate p. But computationally it is quite demanding, particularly if p is large, which may happen in practice quite often. Estimation of p for model (1.4) seems to be an open problem for which we still do not have any satisfactory solution.

The main aim of this Springer brief is to provide a comprehensive review of different aspects of this problem mainly from a statistician's perspective, which is not available in the literature. It is observed that several related models are also available in the literature. We try to provide a brief account on those different models. Interestingly, natural two-dimensional (2D) extension of this model has several applications in texture analysis and in spectrography. We provide a brief review on 2D and three-dimensional (3D) models also. For better understanding of the different procedures discussed in this monograph, we present some real data analysis. Finally we present some open and challenging problems in these areas.

Rest of the monograph is organized as follows. In Chap. 2, we provide the preliminaries. Different methods of estimation are discussed in Chap. 3. Theoretical properties of the different estimators are presented in Chap. 4. Different order estimation methods are reviewed in Chap. 5. Few real data sets are analyzed in Chap. 6. Multidimensional models are introduced in Chap. 7 and finally we provide several related models in Chap. 8.

References

1. Fisher, R. A. (1929). Tests of significance in harmonic analysis. *Proceedings of the Royal Society London Series A, 125*, 54–59.
2. Brillinger, D. (1987). Fitting cosines: Some procedures and some physical examples. In I. B. MacNeill & G. J. Umphrey (Eds.), *Applied probability, stochastic process and sampling theory* (pp. 75–100). Dordrecht: Reidel.
3. Kay, S. M., & Marple, S. L. (1981). Spectrum analysis—A modern perspective. *Proceedings of the IEEE, 69*, 1380–1419.
4. Stoica, P. (1993). List of references on spectral analysis. *Signal Processing, 31*, 329–340.
5. Prasad, S., Chakraborty, M., & Parthasarathy, H. (1995). The role of statistics in signal processing—A brief review and some emerging trends. *Indian Journal of Pure and Applied Mathematics, 26*, 547–578.
6. Kundu, D. (2002). Estimating parameters of sinusoidal frequency; some recent developments. *National Academy Science Letters, 25*, 53–73.
7. Quinn, B. G., & Hannan, E. J. (2001). *The estimation and tracking of frequency*. New York: Cambridge University Press.
8. Hannan, E. J. (1971). Non-linear time series regression. *Journal of Applied Probability, 8*, 767–780.
9. Walker, A. M. (1971). On the estimation of a harmonic component in a time series with stationary independent residuals. *Biometrika, 58*, 21–36.
10. Kundu, D. (1997). Estimating the number of sinusoids in additive white noise. *Signal Processing, 56*, 103–110.

Chapter 2
Notations and Preliminaries

In this monograph, the scalar quantities are denoted by regular lower or uppercase letters. The lower and uppercase bold typefaces of English alphabets are used for vectors and matrices, and for Greek alphabets it should be clear from the context. For a real matrix $\mathbf{A}$, $\mathbf{A}^T$ denotes the transpose. Similarly, for a complex matrix $\mathbf{A}$, $\mathbf{A}^H$ denotes the complex conjugate transpose. An $n \times n$ diagonal matrix , with diagonal elements, $\lambda_1, \ldots, \lambda_n$, are denoted by $\mathrm{diag}\{\lambda_1, \ldots, \lambda_n\}$. If $\mathbf{A}$ is a real or complex square matrix, the projection matrix on the column space of $\mathbf{A}$ is denoted by $\mathbf{P_A} = \mathbf{A}(\mathbf{A}^T\mathbf{A})^{-1}\mathbf{A}^T$ or $\mathbf{P_A} = \mathbf{A}(\mathbf{A}^H\mathbf{A})^{-1}\mathbf{A}^H$ respectively. The following definition and matrix theory results may not be very familiar with the readers and therefore we are providing it for ease of reading.

Definition 2.1 An $n \times n$ matrix $\mathbf{J}$ is called a reflection or exchange matrix if

$$\mathbf{J} = \begin{bmatrix} 0 & 0 & \ldots & 0 & 1 \\ 0 & 0 & \ldots & 1 & 0 \\ \vdots & \vdots & \ddots & \vdots & \vdots \\ 0 & 1 & \ldots & 0 & 0 \\ 1 & 0 & \ldots & 0 & 0 \end{bmatrix}. \tag{2.1}$$

Result 2.1 *(Spectral Decomposition) If $\mathbf{A}$ is an $n \times n$ real symmetric matrix or complex Hermitian matrix, then all the eigenvalues of $\mathbf{A}$ are real and it is possible to find n normalized eigenvectors $v_1, \ldots, v_n$, corresponding to n eigenvalues $\lambda_1, \ldots, \lambda_n$, such that*

$$\mathbf{A} = \sum_{i=1}^{n} \lambda_i v_i v_i^T, \quad or \quad \mathbf{A} = \sum_{i=1}^{n} \lambda_i v_i v_i^H, \tag{2.2}$$

respectively. If $\lambda_i > 0$ for all i, then

D. Kundu and S. Nandi, *Statistical Signal Processing*, SpringerBriefs in Statistics,   7
DOI: 10.1007/978-81-322-0628-6_2, © The Author(s) 2012

$$\mathbf{A}^{-1} = \sum_{i=1}^{n} \frac{1}{\lambda_i} v_i v_i^T, \quad or \quad \mathbf{A}^{-1} = \sum_{i=1}^{n} \frac{1}{\lambda_i} v_i v_i^H \tag{2.3}$$

respectively.

Result 2.2 *(Singular Value Decomposition) If $\mathbf{A}$ is an $n \times m$ real or a complex matrix of rank k, then there exist an $n \times n$ orthogonal matrix $\mathbf{U}$, an $m \times m$ orthogonal matrix $\mathbf{V}$, and an $n \times m$ matrix $\mathbf{\Sigma}$, such that*

$$\mathbf{A} = \mathbf{U \Sigma V}, \tag{2.4}$$

where $\mathbf{\Sigma}$ is defined as

$$\mathbf{\Sigma} = \begin{bmatrix} \mathbf{S} & 0 \\ 0 & 0 \end{bmatrix}, \quad \mathbf{S} = diag\{\sigma_1, \ldots, \sigma_k\},$$

and $\sigma_1^2 \geq \cdots \geq \sigma_k^2 > 0$ are k non-zero eigenvalues of $\mathbf{A}^T \mathbf{A}$ or $\mathbf{A}^H \mathbf{A}$ depending on whether $\mathbf{A}$ is a real or a complex matrix.

Result 2.3 *The following results are used, see Mangulis [1].*

$$\frac{1}{n} \sum_{t=1}^{n} \cos^2(\omega t) = \frac{1}{2} + o\left(\frac{1}{n}\right), \tag{2.5}$$

$$\frac{1}{n} \sum_{t=1}^{n} \sin^2(\omega t) = \frac{1}{2} + o\left(\frac{1}{n}\right), \tag{2.6}$$

$$\frac{1}{n^{k+1}} \sum_{t=1}^{n} t^k \cos(\omega t) \sin(\omega t) = o\left(\frac{1}{n}\right). \tag{2.7}$$

2.1 Prony's Equation

Now, we provide one important result which has been quite extensively used in the Statistical Signal Processing and it is known as Prony's equation. Prony, a Chemical engineer, proposed the following method more than 200 years back in 1795, mainly to estimate the unknown parameters of the real exponential model. It is available in several numerical analysis textbooks, see for example, Froberg [2] or Hildebrand [3]. Prony observed that for arbitrary real constants $\alpha_1, \ldots, \alpha_M$ and for distinct constants $\beta_1, \ldots, \beta_M$, if

$$\mu(t) = \alpha_1 e^{\beta_1 t} + \cdots + \alpha_M e^{\beta_M t}; \quad t = 1, \ldots, n, \tag{2.8}$$

then there exist $(M + 1)$ constants $\{g_0, \ldots, g_M\}$, such that

$$\mathbf{A}\mathbf{g} = \mathbf{0}, \tag{2.9}$$

where

$$\mathbf{A} = \begin{bmatrix} \mu(1) & \cdots & \mu(M+1) \\ \vdots & \ddots & \vdots \\ \mu(n-M) & \cdots & \mu(n) \end{bmatrix}, \quad \mathbf{g} = \begin{bmatrix} g_0 \\ \vdots \\ g_M \end{bmatrix} \quad \text{and} \quad \mathbf{0} = \begin{bmatrix} 0 \\ \vdots \\ 0 \end{bmatrix}. \tag{2.10}$$

Note that without loss of generality we can always put restrictions on $g_0, \ldots, g_M$ such that $\sum_{j=0}^{M} g_j^2 = 1$ and $g_0 > 0$. The sets of linear equations (2.9) is known as Prony's equations. The roots of the following polynomial equation

$$p(x) = g_0 + g_1 x + \cdots + g_M x^M = 0, \tag{2.11}$$

are $e^{\beta_1}, \ldots, e^{\beta_M}$. Therefore, there is a one to one correspondence between $\{\beta_1, \ldots, \beta_M\}$ and $\{g_0, g_1, \ldots, g_M\}$, such that

$$\sum_{j=0}^{M} g_j^2 = 1, \quad g_0 > 0. \tag{2.12}$$

Moreover, $\{g_0, g_1, \ldots, g_M\}$ do not depend on $\{\alpha_1, \ldots, \alpha_M\}$. One natural question is, how to recover $\{\alpha_1, \ldots, \alpha_M\}$ and $\{\beta_1, \ldots, \beta_M\}$ from a given $\mu(1), \ldots, \mu(n)$. It can be done as follows. Note that the rank of matrix $\mathbf{A}$ as defined in (2.10) is M. Therefore, there exists unique $\{g_0, g_1, \ldots, g_M\}$, such that (2.9) and (2.12) hold simultaneously. From that $\{g_0, g_1, \ldots, g_M\}$, using (2.11), $\{\beta_1, \ldots, \beta_M\}$, can be recovered. Now to recover $\{\alpha_1, \ldots, \alpha_M\}$, write (2.8) as

$$\boldsymbol{\mu} = \mathbf{X}\boldsymbol{\alpha}, \tag{2.13}$$

where $\boldsymbol{\mu} = (\mu(1), \ldots, \mu(n))^T$ and $\boldsymbol{\alpha} = (\alpha_1, \ldots, \alpha_M)^T$ are $n \times 1$ and $M \times 1$ vectors, respectively. The $n \times M$ matrix $\mathbf{X}$ is as follows;

$$\mathbf{X} = \begin{bmatrix} e^{\beta_1} & \cdots & e^{\beta_M} \\ \vdots & \ddots & \vdots \\ e^{n\beta_1} & \cdots & e^{n\beta_M} \end{bmatrix}. \tag{2.14}$$

Therefore $\boldsymbol{\alpha} = (\mathbf{X}^T\mathbf{X})^{-1}\mathbf{X}^T\boldsymbol{\mu}$. Note that $\mathbf{X}^T\mathbf{X}$ is a full-rank matrix as $\beta_1, \ldots, \beta_M$ are distinct.

2.2 Undamped Exponential Model

Although Prony observed the relation (2.9) for a real exponential model, the same result is true for a complex exponential model or popularly known as undamped exponential model. A complex exponential model can be expressed as

$$\mu(t) = A_1 e^{i\omega_1 t} + \ldots + A_M e^{i\omega_M t}; \quad t = 1, \ldots, n. \tag{2.15}$$

Here, $A_1, \ldots, A_M$ are complex numbers, $0 < \omega_k < 2\pi$ for $k = 1, \ldots, M$, and $i = \sqrt{-1}$. In this case also there exists $\{g_0, \ldots, g_M\}$, such that they satisfy (2.9). Also the roots of the polynomial equation $p(z) = 0$, as given in (2.11), are $z_1 = e^{i\omega_1}, \ldots, z_M = e^{i\omega_M}$. Observe that

$$|z_1| = \cdots = |z_M| = 1, \quad \bar{z}_k = z_k^{-1}; \quad k = 1, \ldots, M. \tag{2.16}$$

Here $\bar{z}_k$ denotes the complex conjugate of z_k. Define the new polynomial

$$Q(z) = z^{-M} \bar{p}(z) = \bar{g}_0 z^{-M} + \cdots + \bar{g}_M. \tag{2.17}$$

From (2.16), it is clear that $p(z)$ and $Q(z)$ have the same roots. Therefore, we obtain

$$\frac{g_k}{g_M} = \frac{\bar{g}_{M-k}}{\bar{g}_0}; \quad k = 0, \ldots, M, \tag{2.18}$$

by comparing the coefficients of the two polynomials $p(z)$ and $Q(z)$. If we denote

$$b_k = g_k \left(\frac{\bar{g}_0}{g_M} \right)^{-\frac{1}{2}}; \quad k = 0, \ldots, M, \tag{2.19}$$

then

$$b_k = \bar{b}_{M-k}; \quad k = 0, \ldots, M. \tag{2.20}$$

The condition (2.20) is the conjugate symmetric property and can be written as:

$$\mathbf{b} = \mathbf{J}\bar{\mathbf{b}}, \tag{2.21}$$

here $\mathbf{b} = (b_0, \ldots, b_M)^T$ and $\mathbf{J}$ is an exchange matrix as defined in (2.1). From the above discussions, it is immediate that for $\mu(t)$ in (2.15), there exists a vector $\mathbf{g} = (g_0, \ldots, g_M)^T$, such that $\sum_{k=0}^{M} g_k^2 = 1$, which satisfies (2.9), also satisfies

$$\mathbf{g} = \mathbf{J}\bar{\mathbf{g}}. \tag{2.22}$$

2.3 Sum of Sinusoidal Model

Prony's method can also be applied to the sum of sinusoidal model. Suppose $\mu(t)$ can be written as follows:

$$\mu(t) = \sum_{k=1}^{p} [A_k \cos(\omega_k t) + B_k \sin(\omega_k t)]; \quad t = 1, \ldots, n. \qquad (2.23)$$

Here $A_1, \ldots, A_p, B_1, \ldots, B_p$ are real numbers, the frequencies $\omega_1, \ldots, \omega_p$ are distinct and $0 < \omega_k < 2\pi$ for $k = 1, \ldots, p$. Then (2.23) can be written in the form;

$$\mu(t) = \sum_{k=1}^{p} C_k e^{i\omega_k t} + \sum_{k=1}^{p} D_k e^{-i\omega_k t}; \quad t = 1, \ldots, n, \qquad (2.24)$$

here $C_k = (A_k - iB_k)/2$ and $D_k = (A_k + iB_k)/2$. The model (2.24) is in the same form as in (2.15). Therefore, there exists a vector $\mathbf{g} = (g_0, \ldots, g_{2p})$, such that $\sum_{j=0}^{2p} g_j^2 = 1$, which satisfies

$$\begin{bmatrix} \mu(1) & \cdots & \mu(2p+1) \\ \vdots & \ddots & \vdots \\ \mu(n-2p) & \cdots & \mu(n) \end{bmatrix} \begin{bmatrix} g_0 \\ \vdots \\ g_{2p} \end{bmatrix} = \begin{bmatrix} 0 \\ \vdots \\ 0 \end{bmatrix} \qquad (2.25)$$

and satisfies

$$\mathbf{g} = \mathbf{J}\bar{\mathbf{g}}. \qquad (2.26)$$

2.4 Linear Prediction

Observe that $\mu(t)$ as defined in (2.15) also satisfies the following backward linear prediction equation;

$$\begin{bmatrix} \mu(2) & \cdots & \mu(M+1) \\ \vdots & \ddots & \vdots \\ \mu(n-M+1) & \cdots & \mu(n) \end{bmatrix} \begin{bmatrix} d_1 \\ \vdots \\ d_M \end{bmatrix} = - \begin{bmatrix} \mu(1) \\ \vdots \\ \mu(n-M) \end{bmatrix}, \qquad (2.27)$$

clearly, $d_j = b_j/b_0$, for $j = 1, \ldots, M$. It is known as Mth order backward linear prediction equation. In this case the following polynomial equation

$$p(z) = 1 + d_1 z + \cdots + d_M z^M = 0, \tag{2.28}$$

has roots at $e^{i\omega_1}, \ldots, e^{i\omega_M}$.

Along the same manner, it may be noted that $\mu(t)$ as defined in (2.15) also satisfies the following forward linear prediction equation

$$\begin{bmatrix} \mu(M) & \cdots & \mu(1) \\ \vdots & \ddots & \vdots \\ \mu(n-1) & \cdots & \mu(n-M) \end{bmatrix} \begin{bmatrix} a_1 \\ \vdots \\ a_M \end{bmatrix} = - \begin{bmatrix} \mu(M+1) \\ \vdots \\ \mu(n) \end{bmatrix}, \tag{2.29}$$

where $a_1 = b_{M-1}/b_M, a_2 = b_{M-2}/b_M, \ldots, a_M = b_0/b_M$. It is known as Mth order forward linear prediction equation. Moreover, the following polynomial equation

$$p(z) = 1 + a_1 z + \cdots + a_M z^M = 0, \tag{2.30}$$

has roots at $e^{-i\omega_1}, \ldots, e^{-i\omega_M}$.

Consider the case when $\mu(t)$ has form (2.23), then clearly both the polynomial equations (2.28) and (2.30) have roots at $e^{\pm i\omega_1}, \ldots, e^{\pm i\omega_p}$, hence the corresponding coefficients of both the polynomials must be equal. Therefore, in this case the forward backward linear prediction equation can be formed as follows;

$$\begin{bmatrix} \mu(2) & \cdots & \mu(M+1) \\ \vdots & \ddots & \vdots \\ \mu(n-M+1) & \cdots & \mu(n) \\ \mu(M) & \cdots & \mu(1) \\ \vdots & \ddots & \vdots \\ \mu(n-1) & \cdots & \mu(n-M) \end{bmatrix} \begin{bmatrix} d_1 \\ \vdots \\ d_M \end{bmatrix} = - \begin{bmatrix} \mu(1) \\ \vdots \\ \mu(n-M) \\ \mu(M+1) \\ \vdots \\ \mu(n) \end{bmatrix}. \tag{2.31}$$

In the signal processing literature, the linear prediction method has been used quite extensively for different purposes.

2.5 Matrix Pencil

Suppose $\mathbf{A}$ and $\mathbf{B}$ are two $n \times n$, real or complex matrices. The collection of all the matrices $\mathscr{A}$ such that

$$\mathscr{A} = \{\mathbf{C} : \mathbf{C} = \mathbf{A} - \lambda \mathbf{B}, \text{ where } \lambda \text{ is any complex number}\}$$

is called a linear matrix pencil or matrix pencil, see Golub and van Loan [4], and it is denoted by $(\mathbf{A}, \mathbf{B})$. The set of all λ, such that $det(\mathbf{A} - \lambda \mathbf{B}) = 0$, is called the eigenvalues of the matrix pencil $(\mathbf{A}, \mathbf{B})$. The eigenvalues of the matrix pencil $(\mathbf{A}, \mathbf{B})$ can be obtained by solving the general eigenvalue problem of the form

$$\mathbf{Ax} = \lambda \mathbf{Bx}.$$

If $\mathbf{B}^{-1}$ exists, then

$$\mathbf{Ax} = \lambda \mathbf{Bx} \quad \Leftrightarrow \quad \mathbf{B}^{-1}\mathbf{Ax} = \lambda \mathbf{x}.$$

Efficient methods are available to compute the eigenvalues of $(\mathbf{A}, \mathbf{B})$ when $\mathbf{B}^{-1}$ does not exist, or it is nearly a singular matrix, see, for example, Golub and van Loan [4]. The matrix pencil has been used quite extensively in numerical linear algebra. Recently, extensive usage of the matrix pencil method can be found in the spectral estimation method also.

2.6 Stable Distribution: Results

Definition 2.2 A random variable X is said to have a *stable distribution* if there are parameters $0 < \alpha \leq 2, \sigma \geq 0, -1 \leq \beta \leq 1$, and μ real such that its characteristic function has the following form

$$E \exp[it\, X] = \begin{cases} \exp\left\{-\sigma^\alpha |t|^\alpha (1 - i\beta(sign(t))\tan(\frac{\pi\alpha}{2})) + i\mu t\right\} & \text{if } \alpha \neq 1, \\[2mm] \exp\{-\sigma |t|(1 + i\beta\frac{2}{\pi}(sign(t))ln|t|) + i\mu t\} & \text{if } \alpha = 1. \end{cases}$$

The parameter α is called the index of stability or the characteristic exponent and

$$sign(t) = \begin{cases} 1 & \text{if } t > 0 \\ 0 & \text{if } t = 0 \\ -1 & \text{if } t < 0. \end{cases}$$

The parameters σ (scale parameter), β (skewness parameter) and μ are unique (β is irrelevant when $\alpha = 2$) and is denoted by $X \sim S_\alpha(\sigma, \beta, \mu)$.

Definition 2.3 A symmetric (around 0) random variable X is said to have symmetric α stable ($S\alpha S$) distribution, with scale parameter σ, and stability index α, if the characteristic function of the random variable X is

$$E e^{itX} = e^{-\sigma^\alpha |t|^\alpha}. \tag{2.32}$$

We denote the distribution of X as $S\alpha S(\sigma)$. Note that a $S\alpha S$ random variable with $\alpha = 2$ is $\mathcal{N}(0, 2\sigma^2)$ and with $\alpha = 1$, it is a Cauchy random variable, whose density function is

$$f_\sigma(x) = \frac{\sigma}{\pi(x^2 + \sigma^2)}.$$

The $S\alpha S$ distribution is a special case of the general stable distribution with non-zero shift and skewness parameters. For detailed treatments of $S\alpha S$ distribution, the readers are referred to the book of Samorodnitsky and Taqqu [5].

Some Basic Properties:

1. Let X_1 and X_2 be independent random variables with $X_j \sim S_\alpha(\sigma_j, \beta_j, \mu_j)$, $j = 1, 2$, then $X_1 + X_2 \sim S_\alpha(\sigma, \beta, \mu)$ with

$$\sigma = (\sigma_1^\alpha + \sigma_2^\alpha)^{1/\alpha}, \quad \beta = \frac{\beta_1 \sigma_1^\alpha + \beta_2 \sigma_2^\alpha}{\sigma_1^\alpha + \sigma_2^\alpha}, \quad \mu = \mu_1 + \mu_2.$$

2. Let $X \sim S_\alpha(\sigma, \beta, \mu)$ and let a be a real constant. Then $X + a \sim S_\alpha(\sigma, \beta, \mu + a)$.
3. Let $X \sim S_\alpha(\sigma, \beta, \mu)$ and let a be a non-zero real constant. Then

$$aX \sim S_\alpha(|a|\sigma, sign(a)\beta, a\mu) \qquad\qquad\qquad\qquad \text{if } \alpha \neq 1$$
$$aX \sim S_\alpha(|a|\sigma, sign(a)\beta, a\mu - \tfrac{2}{\pi}a(ln|a|)\sigma\beta) \quad \text{if } \alpha = 1$$

.

4. $X \sim S_\alpha(\sigma, \beta, \mu)$ is symmetric if and only if $\beta = 0$ and $\mu = 0$.
5. Let γ be uniform on $(-\pi/2, \pi/2)$ and let W be exponential with mean 1. Assume γ and W independent. Then

$$X = \frac{\sin \alpha\gamma}{(\cos \gamma)^{1/\alpha}} \left(\frac{\cos((1 - \alpha)\gamma)}{W} \right)^{(1-\alpha)/\alpha}$$

is $S_\alpha(1, 0, 0) = S_\alpha(1)$.

Definition 2.4 Let $\mathbf{X} = (X_1, X_2, \ldots, X_d)$ be an α-stable random vector in $\mathbb{R}^d$, then

$$\Phi(\mathbf{t}) = \Phi(t_1, t_2, \ldots, t_d) = E \exp\{i(\mathbf{t}, \mathbf{X})\} = E \exp\{i \sum_{k=1}^{d} t_k X_k\}$$

denote its characteristic function. $\Phi(\mathbf{t})$ is also called the joint characteristic function of the random variables $X_1, X_2, \ldots, X_d$.

Result 2.4 *Let $\mathbf{X}$ be a random vector in $\mathbb{R}^d$.*
(a) If all linear combinations of the random variables $X_1, \ldots, X_d$ are symmetric stable, then $\mathbf{X}$ is a symmetric stable random vector in $\mathbb{R}^d$.
(b) If all linear combinations are stable with index of stability greater than or equal to one, then $\mathbf{X}$ is a stable vector in $\mathbb{R}^d$.

References

1. Mangulis, V. (1965). *Handbook of series for scientists and engineers*. New York: Academic Press.
2. Froberg, C.E. (1969). Introduction to numerical analysis, 2nd-ed., Addision-Wesley Pub. Co., Boston.
3. Hilderband, F. B. (1956). *An introduction to numerical analysis*. New York: McGraw-Hill.
4. Golub, G. and van Loan, C. (1996). Matrix computations, 3-rd. ed., The Johns Hopkins University Press, London.
5. Samorodnitsky, G., & Taqqu, M. S. (1994). *Stable non-Gaussian random processes; stochastic models with infinite variance*. New York: Chapman and Hall.

Chapter 3
Estimation of Frequencies

In this section, we provide different estimation procedures of the frequencies of a periodic signal. We consider the following sum of sinusoidal model;

$$y(t) = \sum_{k=1}^{p} (A_k \cos(\omega_k t) + B_k \sin(\omega_k t)) + X(t); \quad t = 1, \ldots, n. \qquad (3.1)$$

Here A_k, B_k, ω_k for $k = 1, \ldots, p$ are unknown. In this chapter, p is assumed to be known, later in Chap. 5, we provide different estimation methods of p. The error component $X(t)$ has mean zero and finite variance, and it can have either one of the following forms:

Assumption 3.1 $\{X(t); t = 1, \ldots, n\}$ are i.i.d. random variables with $E(X(t)) = 0$ and $V(X(t)) = \sigma^2$.

Assumption 3.2 $\{X(t)\}$ is a stationary linear process with the following form

$$X(t) = \sum_{j=0}^{\infty} a(j)e(t - j), \qquad (3.2)$$

where $\{e(t); t = 1, 2, \ldots\}$ are i.i.d. random variables with $E(e(t)) = 0$, $V(e(t)) = \sigma^2$, and $\sum_{j=0}^{\infty} |a(j)| < \infty$.

All the available methods have been used under both these assumptions. Although theoretical properties of these estimators obtained by different methods may not be available under both these assumptions.

In this chapter, we mainly discuss different estimation procedures, while the properties of these estimators are discussed in Chap. 4. Most of the methods deal with the estimation of frequencies. If the frequencies are known, model (3.1) can be treated as a linear regression model, therefore the linear parameters A_k and B_k for $k = 1, \ldots, p$

D. Kundu and S. Nandi, *Statistical Signal Processing*, SpringerBriefs in Statistics,
DOI: 10.1007/978-81-322-0628-6_3, © The Author(s) 2012

can be estimated using a simple least squares or generalized least squares method depending on the error structure. Similar methods can be adopted even when the frequencies are unknown.

3.1 ALSEs and PEs

In this section, first we assume that $p = 1$, later we provide the method for general p. Under Assumption 3.1, the most intuitive estimators are the LSEs, that is, $\widehat{A}$, $\widehat{B}$, and $\widehat{\omega}$, which minimize

$$Q(A, B, \omega) = \sum_{t=1}^{n} (y(t) - A\cos(\omega t) - B\sin(\omega t))^2. \tag{3.3}$$

Finding the LSEs is a non-linear optimization problem, and it is well known to be a numerically difficult problem. The standard algorithm like Newton–Raphson, Gauss–Newton or their variants may be used to minimize (3.3). Often it is observed that the standard algorithms may not converge, even when the iterative process starts from a very good starting value. It is observed by Rice and Rosenblatt [1], that the least squares surface has several local minima near the true parameter value, and due to this reason most of the iterative procedures even when they converge, often converge to a local minimum rather than the global minimum.

Therefore, even if it is known that LSEs are the most efficient estimators, finding LSEs is a numerically challenging problem. Due to this reason, extensive work has been done in the statistical signal processing literature to find estimators that perform like the LSEs.

First, we provide the most popular method, which has been used in practice to compute the frequency estimator. For $p = 1$, model (3.1) can be written as

$$\mathbf{Y} = \mathbf{Z}(\omega)\theta + \mathbf{e}, \tag{3.4}$$

where

$$\mathbf{Y} = \begin{bmatrix} y(1) \\ \vdots \\ y(n) \end{bmatrix}, \quad \mathbf{Z}(\omega) = \begin{bmatrix} \cos(\omega) & \sin(\omega) \\ \vdots & \vdots \\ \cos(n\omega) & \sin(n\omega) \end{bmatrix}, \quad \theta = \begin{bmatrix} A \\ B \end{bmatrix}, \quad \mathbf{e} = \begin{bmatrix} X(1) \\ \vdots \\ X(n) \end{bmatrix}. \tag{3.5}$$

Therefore, for a given ω, the LSEs of A and B can be obtained as

$$\widehat{\theta}(\omega) = \begin{bmatrix} \widehat{A}(\omega) \\ \widehat{B}(\omega) \end{bmatrix} = \left(\mathbf{Z}^T(\omega)\mathbf{Z}(\omega)\right)^{-1} \mathbf{Z}^T(\omega)\mathbf{Y}. \tag{3.6}$$

Using (2.5)–(2.7), (3.6) can be written as

$$\begin{bmatrix} \widehat{A}(\omega) \\ \widehat{B}(\omega) \end{bmatrix} = \begin{bmatrix} 2\sum_{t=1}^{n} y(t)\cos(\omega t)/n \\ 2\sum_{t=1}^{n} y(t)\sin(\omega t)/n \end{bmatrix}. \tag{3.7}$$

Substituting $\widehat{A}(\omega)$ and $\widehat{B}(\omega)$ in (3.3), we obtain

$$\frac{1}{n} Q\left(\widehat{A}(\omega), \widehat{B}(\omega), \omega\right) = \frac{1}{n}\mathbf{Y}^T\mathbf{Y} - \frac{1}{n}\mathbf{Y}^T\mathbf{Z}(\omega)\mathbf{Z}^T(\omega)\mathbf{Y} + o(1). \tag{3.8}$$

Therefore, $\widehat{\omega}$ which minimizes $Q(\widehat{A}(\omega), \widehat{B}(\omega), \omega)/n$ is equivalent to $\widetilde{\omega}$, which maximizes

$$I(\omega) = \frac{1}{n}\mathbf{Y}^T\mathbf{Z}(\omega)\mathbf{Z}^T(\omega)\mathbf{Y} = \frac{1}{n}\left\{ \left(\sum_{t=1}^{n} y(t)\cos(\omega t)\right)^2 + \left(\sum_{t=1}^{n} y(t)\sin(\omega t)\right)^2 \right\}$$

in the sense $\widehat{\omega} - \widetilde{\omega} \overset{a.e.}{\to} 0$. The estimator of ω, which is obtained by maximizing $I(\omega)$ for $0 \leq \omega \leq \pi$, is known as the ALSE of ω.

The maximization of $I(\omega)$ can be performed by some standard algorithm like Newton–Raphson or Gauss–Newton method, although computation of the ALSEs has the same type of problems as the LSEs. In practice instead of maximizing $I(\omega)$ for $0 < \omega < \pi$, it is maximized at the Fourier frequencies, namely at the points $2\pi j/n; 0 \leq j < [n/2]$. Therefore, $\widetilde{\widetilde{\omega}} = 2\pi j_0/n$ is an estimator of ω, where

$$I(\omega_{j_0}) > I(\omega_k), \quad \text{for} \quad k = 1, \ldots, [n/2], k \neq j_0.$$

It is also known as the PE of ω. Although it is not an efficient estimator, it is being used extensively as an initial guess of any iterative procedure to compute an efficient estimator of the frequency.

The method can be easily extended for the model when $p > 1$. The main idea is to remove the effect of the first component from the signal $\{y(t)\}$ and repeat the whole procedure. The details are explained in Sect. 3.12.

3.2 EVLP

The equivariance linear prediction (EVLP) method was suggested by Bai et al. [2] for estimating the frequencies of model (3.1), under error Assumption 3.1. It mainly uses the fact that in the absence of $\{X(t); t = 1, \cdots, n\}$, $\{y(t); t = 1, \cdots, n\}$ satisfy (2.25). The idea behind the EVLP method is as follows: Consider an $(n - 2p) \times n$ data matrix $\mathbf{Y}_D$ as

$$\mathbf{Y}_D = \begin{bmatrix} y(1) & \cdots & y(2p+1) \\ \vdots & \ddots & \vdots \\ y(n-2p) & \cdots & y(n) \end{bmatrix}. \tag{3.9}$$

If $\{X(t)\}$ is absent, $\text{Rank}(\mathbf{Y}_D) = \text{Rank}(\mathbf{Y}_D^T \mathbf{Y}_D/n) = 2p$. It implies that the symmetric matrix $(\mathbf{Y}_D^T \mathbf{Y}_D/n)$ has an eigenvalue zero with multiplicity one. Therefore, there exists an eigenvector $\mathbf{g} = (g_0, \ldots, g_{2p})$ which corresponds to the zero eigenvalue, such that $\sum_{j=0}^{2p} g_j^2 = 1$, and the polynomial equation

$$p(x) = g_0 + g_1 x + \cdots + g_{2p} x^{2p} = 0 \tag{3.10}$$

has roots at $e^{\pm i\omega_1}, \ldots, e^{\pm i\omega_p}$.

Using this idea, Bai et al. [2] proposed that when $\{X(t)\}$ satisfies error Assumption 3.1, from the symmetric matrix $(\mathbf{Y}_D^T \mathbf{Y}_D/n)$ obtain the normalized eigenvector $\widehat{\mathbf{g}} = (\widehat{g}_0, \ldots, \widehat{g}_{2p})$, such that $\sum_{j=0}^{2p} \widehat{g}_j^2 = 1$ corresponds to the minimum eigenvalue. Form the polynomial equation

$$\widehat{p}(x) = \widehat{g}_0 + \widehat{g}_1(x) + \cdots + \widehat{g}_{2p} x^{2p} = 0, \tag{3.11}$$

and obtain the estimates of $\omega_1, \cdots, \omega_p$ from these estimated roots.

It has been shown by Bai et al. [2] that as $n \to \infty$, EVLP method provides consistent estimators of the unknown frequencies. It is interesting that although EVLP frequency estimators are consistent, the corresponding linear parameter estimators obtained by the least squares method as mentioned before are not consistent estimators. Moreover, it has been observed in the extensive simulation studies by Bai et al. [3] that the performance of the EVLP estimators is not very satisfactory for small sample sizes.

3.3 MFBLP

In the signal processing literature, the forward linear prediction method or backward linear prediction, see Sect. 2.4, has been used to estimate the frequencies of the sinusoidal signals. It has been observed by Kumaresan [4] using extensive simulation studies that the pth order linear prediction method does not work very well in estimating the frequencies in the presence of noise, particularly when the two frequencies are very close to each other.

Due to this reason, Kumaresan [4], see also Tufts and Kumaresan [5], used the extended order forward backward linear prediction method, and call it as the modified forward backward linear prediction (MFBLP) method, and it can be described as follows. Choose an L, such that $n - 2p \geq L > 2p$, and set up the Lth order backward and forward linear prediction equations as follows;

$$
\begin{bmatrix}
y(L) & \cdots & y(1) \\
\vdots & \ddots & \vdots \\
y(n-1) & \cdots & y(n-L) \\
y(2) & \cdots & y(L+1) \\
\vdots & \ddots & \vdots \\
y(n-L+1) & \cdots & y(n)
\end{bmatrix}
\begin{bmatrix}
b_1 \\
\vdots \\
b_L
\end{bmatrix}
= -
\begin{bmatrix}
y(L+1) \\
\vdots \\
y(n) \\
y(1) \\
\vdots \\
y(n-L)
\end{bmatrix} .
\tag{3.12}
$$

It has been shown by Kumaresan [4] that in the noiseless case, $\{y(t)\}$ satisfies (3.12), and out of the L roots of the polynomial equation

$$
p(x) = x^L + b_1 x^{L-1} + \cdots + b_L = 0,
\tag{3.13}
$$

$2p$ roots are of the form $e^{\pm i\omega_k}$, and the rest of the $L - 2p$ roots are of the modulus not equal to one. Observe that (3.12) can be written as

$$
\mathbf{Ab} = -\mathbf{h}.
\tag{3.14}
$$

Tufts and Kumaresan [5] suggested using the truncated singular value decomposition solution of the vector $\mathbf{b}$ by setting the smaller singular values of the matrix $\mathbf{A}$ equal to zero. Therefore, if the singular value decomposition of $\mathbf{A}$ is as given in (2.4) where $\mathbf{v}_k$ and $\mathbf{u}_k$ for $k = 1, \cdots, 2p$ are the eigenvectors of $\mathbf{A}^T \mathbf{A}$ and $\mathbf{A}\mathbf{A}^T$, respectively, $\sigma_1^2 \geq \cdots \sigma_{2p}^2 > 0$ are the $2p$ non-zero eigenvalues of $\mathbf{A}^T \mathbf{A}$, then the solution $\widehat{\mathbf{b}}$ of the system of Eq. (3.14) becomes

$$
\widehat{\mathbf{b}} = -\sum_{k=1}^{2p} \frac{1}{\sigma_k} \left[\mathbf{u}_k^T \mathbf{h} \right] \mathbf{v}_k.
\tag{3.15}
$$

The effect of using the truncated singular value decomposition is to increase the signal-to-noise ratio in the noisy data, prior to obtaining the solution vector $\widehat{\mathbf{b}}$. Once $\widehat{\mathbf{b}}$ is obtained, get the L roots of the L-degree polynomial

$$
p(x) = x^L + \widehat{b}_1 x^{L-1} + \cdots + \widehat{b}_L,
\tag{3.16}
$$

and choose $2p$ roots, which are closest to one in absolute value. It is expected, if at least the variance is small, that they form p conjugate pairs and from there the frequencies can be easily estimated.

It is observed by extensive simulation studies by Kumaresan [4] that the MFBLP performs very well if $L \approx 2n/3$, and the error variance is not too large. The main computation involved in this case is the computation of the singular value decomposition of $\mathbf{A}$ and then root findings of an L degree polynomial. Although, the

MFBLP performs very well for small sizes, it has been pointed out by Rao [6] that MFBLP estimators are not consistent.

3.4 NSD

The noise space decomposition (NSD) method has been proposed by Kundu and Mitra [7], see also Kundu and Mitra [8] in this respect. The basic idea behind the NSD method can be described as follows. Consider the following $(n - L) \times (L + 1)$ matrix $\mathbf{A}$, where

$$
\mathbf{A} = \begin{bmatrix} \mu(1) & \cdots & \mu(L+1) \\ \vdots & \ddots & \vdots \\ \mu(n-L) & \cdots & \mu(n) \end{bmatrix},
\tag{3.17}
$$

for any integer L, such that $2p \leq L \leq n - 2p$, and $\mu(t)$ is same as defined in (2.23). Let the spectral decomposition of $\mathbf{A}^T \mathbf{A}/n$ be

$$
\frac{1}{n} \mathbf{A}^T \mathbf{A} = \sum_{i=1}^{L+1} \sigma_i^2 \mathbf{u}_i \mathbf{u}_i^T,
\tag{3.18}
$$

where $\sigma_1^2 \geq \cdots \geq \sigma_{L+1}^2$ are the eigenvalues of $\mathbf{A}^T \mathbf{A}/n$ and $\mathbf{u}_1, \ldots, \mathbf{u}_{L+1}$ are the corresponding orthonormal eigenvectors. Since matrix $\mathbf{A}$ is of rank $2p$,

$$
\sigma_{2p+1}^2 = \cdots = \sigma_{L+1}^2 = 0,
$$

and the null space spanned by the columns of matrix $\mathbf{A}^T \mathbf{A}$ is of rank $L + 1 - 2p$. Using Prony's equations one obtains

$$
\mathbf{AB} = \mathbf{0},
$$

where $\mathbf{B}$ is an $(L + 1) \times (L + 1 - 2p)$ matrix of rank $(L + 1 - 2p)$ as follows;

$$
\mathbf{B} = \begin{bmatrix} g_0 & 0 & \cdots & 0 \\ g_1 & g_0 & \cdots & 0 \\ \vdots & \vdots & \ddots & \vdots \\ g_{2p} & g_{2p-1} & \cdots & 0 \\ 0 & g_{2p} & \cdots & g_0 \\ 0 & 0 & \ddots & g_1 \\ \vdots & \vdots & \cdots & \vdots \\ 0 & 0 & \cdots & g_{2p} \end{bmatrix},
\tag{3.19}
$$

and $g_0, \ldots, g_{2p}$ are same as defined before. Moreover, the space spanned by the columns of $\mathbf{B}$ is the null space spanned by the columns of matrix $\mathbf{A}^T \mathbf{A}$.

Consider the following $(n - L) \times (L + 1)$ data matrix $\widetilde{\mathbf{A}}$ as follows;

$$\widetilde{\mathbf{A}} = \begin{bmatrix} y(1) & \cdots & y(L+1) \\ \vdots & \ddots & \vdots \\ y(n-L) & \cdots & y(n) \end{bmatrix}.$$

Let the spectral decomposition of $\widetilde{\mathbf{A}}^T \widetilde{\mathbf{A}}/n$ be

$$\frac{1}{n}\widetilde{\mathbf{A}}^T \widetilde{\mathbf{A}} = \sum_{i=1}^{L+1} \widetilde{\sigma}_i^2 \widetilde{\mathbf{u}}_i \widetilde{\mathbf{u}}_i^T,$$

where $\widetilde{\sigma}_1^2 > \cdots > \widetilde{\sigma}_{L+1}^2$ are ordered eigenvalues of $\widetilde{\mathbf{A}}^T \widetilde{\mathbf{A}}/n$ and $\widetilde{\mathbf{u}}_1, \ldots, \widetilde{\mathbf{u}}_{L+1}$ are orthonormal eigenvectors corresponding to $\widetilde{\sigma}_1^2, \ldots, \widetilde{\sigma}_{L+1}$, respectively. Construct $(L + 1) \times (L + 1 - 2p)$ matrix $\mathbf{C}$ as

$$\mathbf{C} = \begin{bmatrix} \widetilde{\mathbf{u}}_{2p+1} \vdots \cdots \vdots \widetilde{\mathbf{u}}_{L+1} \end{bmatrix}.$$

Partition matrix $\mathbf{C}$ as

$$\mathbf{C}^T = \begin{bmatrix} \mathbf{C}_{1k}^T : \mathbf{C}_{2k}^T : \mathbf{C}_{3k}^T \end{bmatrix},$$

for $k = 0, 1, \ldots, L - 2p$, where $\mathbf{C}_{1k}^T$, $\mathbf{C}_{2k}^T$, and $\mathbf{C}_{3k}^T$ are of orders $(L + 1 - 2p) \times k$, $(L + 1 - 2p) \times (2p + 1)$, and $(L + 1 - 2p) \times (L - k + 2p)$, respectively. Find an $(L + 1 - 2p)$ vector $\mathbf{x}_k$, such that

$$\begin{bmatrix} \mathbf{C}_{1k}^T \\ \mathbf{C}_{3k}^T \end{bmatrix} \mathbf{x}_k = \mathbf{0}.$$

Denote the vectors for $k = 0, 1, \ldots, L - 2p$

$$\mathbf{b}^k = \mathbf{C}_{2k}^T \mathbf{x}_k,$$

and consider the vector $\mathbf{b}$, the average of the vectors $\mathbf{b}^0, \ldots, \mathbf{b}^{L-2p}$, that is

$$\mathbf{b} = \frac{1}{L+1-2p} \sum_{k=0}^{L-2p} \mathbf{b}^k = \begin{bmatrix} \widehat{g}_0, \widehat{g}_1, \ldots, \widehat{g}_{2p} \end{bmatrix}.$$

Construct the polynomial equation

$$\widehat{g}_0 + \widehat{g}_1 x + \cdots + \widehat{g}_{2p} x^{2p} = 0, \tag{3.20}$$

and obtain the estimates of the frequencies from complex conjugate roots of (3.20).

It has been shown by Kundu and Mitra [8] that the NSD estimators are strongly consistent, although asymptotic distribution of the NSD estimators has not yet been established. It is observed by extensive simulation studies that the performance of the NSD estimators are very good and it provides the best performance when $L \approx n/3$. The main computation of the NSD estimators involves computation of the singular value decomposition of an $(L + 1) \times (L + 1)$ matrix, and the root findings of a $2p$ degree polynomial equation.

3.5 ESPRIT

Estimation of signal parameters via rotational invariance technique (ESPRIT) was first proposed by Roy [9] in his Ph.D. thesis, see also Roy and Kailath [10], which is based on the generalized eigenvalue-based method. The basic idea comes from Prony's system of homogeneous equations. For a given L, when $2p < L < n - 2p$, construct the two data matrices $\mathbf{A}$ and $\mathbf{B}$ both of the order $(n - L) \times L$ as given below;

$$\mathbf{A} = \begin{bmatrix} y(1) & \cdots & y(L) \\ \vdots & \ddots & \vdots \\ y(n - L) & \cdots & y(n - 1) \end{bmatrix}, \quad \mathbf{B} = \begin{bmatrix} y(2) & \cdots & y(L + 1) \\ \vdots & \ddots & \vdots \\ y(n - L + 1) & \cdots & y(n) \end{bmatrix}.$$

If, $\mathbf{C}_1 = (\mathbf{A}^T \mathbf{A} - \sigma^2 \mathbf{I})$ and $\mathbf{C}_2 = \mathbf{B}^T \mathbf{A} - \sigma^2 \mathbf{K}$, where $\mathbf{I}$ is the identity matrix of order $L \times L$ and $\mathbf{K}$ is an $L \times L$ matrix with ones along the first lower diagonal off the major diagonal and zeros elsewhere. Consider the matrix pencil $\mathbf{C}_1 - \gamma \mathbf{C}_2$.

It has been shown, see Pillai [11] for details, that out of the L eigenvalues of the matrix pencil $\mathbf{C}_1 - \gamma \mathbf{C}_2$, $2p$ non-zero eigenvalues are of the form $e^{\pm i\omega_k}, k = 1, \ldots, p$. Therefore, from those $2p$ non-zero eigenvalues the unknown frequencies $\omega_1, \ldots, \omega_p$ can be estimated. It is further observed that if $\sigma^2 = 0$, then $L - 2p$ eigenvalues of the matrix pencil $\mathbf{C}_1 - \gamma \mathbf{C}_2$ are zero.

The following problems are observed to implement the ESPRIT in practice. Note that both the matrices $\mathbf{C}_1$ and $\mathbf{C}_2$ involve σ^2, which is unknown. If σ^2 is very small it may be ignored, otherwise it needs to be estimated, or some prior knowledge may be used. Another problem is to separate $2p$ non-zero eigenvalues from the total of L eigenvalues. Again if σ^2 is small, it may not be much of a problem, but for large σ^2 separation of non-zero eigenvalues from the zero eigenvalues may not be a trivial issue. The major computational issue is to compute the eigenvalues of the matrix pencil $\mathbf{C}_1 - \gamma \mathbf{C}_2$, and the problem is quite ill conditioned if σ^2 is small. In this case, the matrices $\mathbf{A}$ and $\mathbf{B}$ both are nearly singular matrices. To avoid both these issues, the following method has been suggested.

3.6 TLS-ESPRIT

Total least squares ESPRIT (TLS-ESPRIT) has been proposed by Roy and Kailath [10] mainly to overcome some of the problems involved in implementing the ESPRIT algorithm in practice. Using the same notation as in Sect. 3.5, construct the following $2L \times 2L$ matrices $\mathbf{R}$ and $\mathbf{\Sigma}$ as follows;

$$\mathbf{R} = \begin{bmatrix} \mathbf{A}^T \\ \mathbf{B}^T \end{bmatrix} [\mathbf{A} \ \mathbf{B}] \quad \text{and} \quad \mathbf{\Sigma} = \begin{bmatrix} \mathbf{I} & \mathbf{K} \\ \mathbf{K}^T & \mathbf{I} \end{bmatrix}.$$

Let $\mathbf{e}_1, \ldots, \mathbf{e}_{2p}$ be $2p$ orthonormal eigenvectors of the matrix pencil $(\mathbf{R} - \gamma \mathbf{\Sigma})$ corresponding to the largest $2p$ eigenvalues. Now construct the two $L \times 2p$ matrices $\mathbf{E}_1$ and $\mathbf{E}_2$ from $\mathbf{e}_1, \ldots, \mathbf{e}_{2p}$ as

$$[\mathbf{e}_1 : \cdots : \mathbf{e}_{2p}] = \begin{bmatrix} \mathbf{E}_1 \\ \mathbf{E}_2 \end{bmatrix}$$

and then obtain the unique $4p \times 2p$ matrix $\mathbf{W}$ and the two $2p \times 2p$ matrices $\mathbf{W}_1$ and $\mathbf{W}_2$ as follows

$$[\mathbf{E}_1 : \mathbf{E}_2]\mathbf{W} = \mathbf{0} \quad \text{and} \quad \mathbf{W} = \begin{bmatrix} \mathbf{W}_1 \\ \mathbf{W}_2 \end{bmatrix}.$$

Finally obtain the $2p$ eigenvalues of $-\mathbf{W}_1\mathbf{W}_2^{-1}$. Again, it has been shown, see Pillai [11], that in the noise less case, the above $2p$ eigenvalues are of the form $e^{\pm i\omega_k}$ for $k = 1, \ldots, p$. Therefore, the frequencies can be estimated from the eigenvalues of the matrix $-\mathbf{W}_1\mathbf{W}_2^{-1}$.

It is known that the performance of TLS-ESPRIT is very good, and it is much better than the ESPRIT method. The performance of both the methods depends on the values of L. In this case also, the main computation involves the computation of the eigenvalues and eigenvectors of the $L \times L$ matrix pencil $(\mathbf{R} - \gamma \mathbf{\Sigma})$. Although, the performance of the TLS-ESPRIT is very good, the consistency property of TLS-ESPRIT or ESPRIT has not yet been established.

3.7 Quinn's Method

Quinn [12] proposed this method in estimating the frequency of model (3.1) when $p = 1$. It can be easily extended for general p, see, for example Kundu and Mitra [8]. Quinn's method can be applied in the presence of error Assumption 3.2. The method is based on the interpolation of the Fourier coefficients, and using the fact that the PE has the convergence rate $O(1/n)$.

The method can be described as follows: Let

$$Z(j) = \sum_{t=1}^{n} y(t)e^{-i2\pi jt/n}; \quad j = 1, \dots, n.$$

Algorithm 3.1

- Step 1: Let $\widetilde{\omega}$ be the maximizer of $|Z(j)|^2$, for $1 \le j \le n$.
- Step 2: Let $\widehat{\alpha}_1 = Re\{Z(\widetilde{\omega} - 1)/Z(\widetilde{\omega})\}$, $\widehat{\alpha}_2 = Re\{Z(\widetilde{\omega} + 1)/Z(\widetilde{\omega})\}$, and $\widehat{\delta}_1 = \widehat{\alpha}_1/(1 - \widehat{\alpha}_1)$, $\widehat{\delta}_2 = -\widehat{\alpha}_2/(1 - \widehat{\alpha}_2)$. If $\widehat{\delta}_1 > 0$ and $\widehat{\delta}_2 > 0$, put $\widehat{\delta} = \widehat{\delta}_2$, otherwise put $\widehat{\delta} = \widehat{\delta}_1$.
- Step 3: Estimate ω by $\widehat{\omega} = 2\pi(\widetilde{\omega} + \widehat{\delta})/n$.

Computationally Quinn's method is very easy to implement. It is observed that Quinn's method produces consistent estimator of the frequency, and the asymptotic mean squared error of the frequency estimator is of the order $O(1/n^3)$. Although Quinn's method has been proposed for one component only, the method can be easily extended for the model when $p > 1$. The details are explained in Sect. 3.12.

3.8 IQML

The iterative quadratic maximum likelihood (IQML) method was proposed by Bresler and Macovski [13], and this is the first special purpose algorithm, which has been used to compute the LSEs of the unknown parameters of model (3.1). It is well known that in the presence of i.i.d. additive normal errors, the LSEs become the MLEs also. Re-write model (3.1) as follows:

$$\mathbf{Y} = \mathbf{Z}(\boldsymbol{\omega})\boldsymbol{\theta} + \mathbf{e}, \tag{3.21}$$

where

$$\mathbf{Y} = \begin{bmatrix} y(1) \\ \vdots \\ y(n) \end{bmatrix}, \mathbf{Z}(\boldsymbol{\omega}) = \begin{bmatrix} \cos(\omega_1) & \sin(\omega_1) & \cdots & \cos(\omega_p) & \sin(\omega_p) \\ \vdots & \vdots & \ddots & \vdots & \vdots \\ \cos(n\omega_1) & \sin(n\omega_1) & \cdots & \cos(n\omega_p) & \sin(n\omega_p) \end{bmatrix},$$

$$\boldsymbol{\theta}^T = [A_1, B_1, \dots, A_p, B_p], \mathbf{e}^T = [e_1, \dots, e_n], \boldsymbol{\omega}^T = (\omega_1, \dots, \omega_p).$$

Therefore, the LSEs of the unknown parameters can be obtained by minimizing

$$Q(\boldsymbol{\omega}, \boldsymbol{\theta}) = (\mathbf{Y} - \mathbf{Z}(\boldsymbol{\omega})\boldsymbol{\theta})^T (\mathbf{Y} - \mathbf{Z}(\boldsymbol{\omega})\boldsymbol{\theta}) \tag{3.22}$$

with respect to $\boldsymbol{\omega}$ and $\boldsymbol{\theta}$. Therefore, for a given $\boldsymbol{\omega}$, the LSE of $\boldsymbol{\theta}$ can be obtained as

$$\widehat{\theta}(\omega) = (\mathbf{Z}(\omega)^T \mathbf{Z}(\omega))^{-1} \mathbf{Z}(\omega)^T \mathbf{Y}. \tag{3.23}$$

Substituting back (3.23) in (3.22) we obtain

$$R(\omega) = Q(\widehat{\theta}(\omega), \omega) = \mathbf{Y}^T (\mathbf{I} - \mathbf{P_Z})\mathbf{Y}, \tag{3.24}$$

here $\mathbf{P_Z} = \mathbf{Z}(\omega)(\mathbf{Z}(\omega)^T \mathbf{Z}(\omega))^{-1}\mathbf{Z}(\omega)^T$ is the projection matrix on the space spanned by the columns of $\mathbf{Z}(\omega)$. Note that $\mathbf{I} - \mathbf{P_Z} = \mathbf{P_B}$, where the matrix $\mathbf{B} = \mathbf{B}(\mathbf{g})$ is same as defined in (3.19), and $\mathbf{P_B} = \mathbf{B}(\mathbf{B}^T\mathbf{B})^{-1}\mathbf{B}^T$ is the projection matrix orthogonal to $\mathbf{P_Z}$. The IQML method mainly suggests how to minimize $\mathbf{Y}^T\mathbf{P_B}\mathbf{Y}$ with respect to the unknown vector $\mathbf{g} = (g_0, \ldots, g_{2p})^T$, which is equivalent to minimize (3.24) with respect to the unknown parameter vector ω. First observe that

$$\mathbf{Y}^T \mathbf{P_B} \mathbf{Y} = \mathbf{g}^T \mathbf{Y}_D^T (\mathbf{B}^T \mathbf{B})^{-1} \mathbf{Y}_D \mathbf{g}, \tag{3.25}$$

where $\mathbf{Y}_D$ is an $(n - 2p) \times (2p + 1)$ matrix as defined in (3.9). The following algorithm has been suggested by Bresler and Macovski [13] to minimize $\mathbf{Y}^T\mathbf{P_B}\mathbf{Y} = \mathbf{g}^T\mathbf{Y}_D^T(\mathbf{B}^T\mathbf{B})^{-1}\mathbf{Y}_D\mathbf{g}$.

Algorithm 3.2

1. Suppose at the kth step the value of the vector $\mathbf{g}$ is $\mathbf{g}_{(k)}$.
2. Compute matrix $\mathbf{C}_{(k)} = \mathbf{Y}_D^T(\mathbf{B}_{(k)}^T\mathbf{B}_{(k)})^{-1}\mathbf{Y}_D$, here $\mathbf{B}_{(k)}$ is obtained by replacing $\mathbf{g}$ with $\mathbf{g}_{(k)}$ in matrix $\mathbf{B}$ given in (3.19).
3. Solve the quadratic optimization problem

$$\min_{\mathbf{x}:||\mathbf{x}||=1} \mathbf{x}^T \mathbf{C}_{(k)} \mathbf{x},$$

and suppose the solution is $\mathbf{g}_{(k+1)}$.
4. Check the convergence whether $|\mathbf{g}_{(k+1)} - \mathbf{g}_{(k)}| < \varepsilon$, where ε is some pre assigned value. If the convergence is met, go to step [5], otherwise go to step [1].
5. Obtain the estimate of ω from the estimate of $\mathbf{g}$.

Although, no proof of convergence is available for the above algorithm, it works quite well in practice.

3.9 Modified Prony Algorithm

Modified Prony algorithm was proposed by Kundu [14], which also involves the minimization of

$$\Psi(\mathbf{g}) = \mathbf{Y}^T \mathbf{P_B} \mathbf{Y} \tag{3.26}$$

with respect to the vector $\mathbf{g}$, where the matrix $\mathbf{B}(\mathbf{g})$ and the projection matrix $\mathbf{P_B}$ are same as defined in Sect. 3.8. Now observe that $\Psi(\mathbf{g})$ is invariant under scalar multiplication, that is

$$\Psi(\mathbf{g}) = \Psi(c\mathbf{g}),$$

for any constant $c \in \mathbb{R}$. Therefore,

$$\min_{\mathbf{g}} \Psi(\mathbf{g}) = \min_{\mathbf{g}:\mathbf{g}^T\mathbf{g}=1} \Psi(\mathbf{g}).$$

To minimize $\Psi(\mathbf{g})$ with respect to $\mathbf{g}$, differentiate $\mathbf{g}$ with respect to different components of $\mathbf{g}$ and equating them to zero lead to solving the following non-linear equation

$$\mathbf{C}(\mathbf{g})\mathbf{g} = \mathbf{0}; \quad \mathbf{g}^T\mathbf{g} = 1, \tag{3.27}$$

where $\mathbf{C} = \mathbf{C}(\mathbf{g})$ is a $(2p+1) \times (2p+1)$ symmetric matrix whose (i, j)th element, for $i, j = 1, \ldots, 2p+1$, is given by

$$c_{ij} = \mathbf{Y}^T \mathbf{B}_i (\mathbf{B}^T\mathbf{B})^{-1} \mathbf{B}_j^T \mathbf{Y} - \mathbf{Y}^T \mathbf{B}(\mathbf{B}^T\mathbf{B})^{-1} \mathbf{B}_j^T \mathbf{B}_i (\mathbf{B}^T\mathbf{B})^{-1} \mathbf{B}^T \mathbf{Y}.$$

Here the elements of matrix $\mathbf{B}_j^T$, for $j = 1, \ldots, 2p+1$, are only zero and ones, such that

$$\mathbf{B}(\mathbf{g}) = \sum_{j=0}^{2p} g_j \mathbf{B}_j.$$

The problem (3.27) is a non-linear eigenvalue problem and the following iterative scheme has been suggested by Kundu [14] to solve the set of non-linear equations;

$$(\mathbf{C}(\mathbf{g}^{(k)}) - \lambda^{(k+1)}\mathbf{I})\mathbf{g}^{(k+1)} = \mathbf{0}; \quad \mathbf{g}^{(k+1)T}\mathbf{g}^{(k+1)} = 1. \tag{3.28}$$

Here $\mathbf{g}^{(k)}$ denotes the kth iterate of the above iterative process, and $\lambda^{(k+1)}$ is the eigenvalue of $\mathbf{C}(\mathbf{g}^{(k)})$, which is closest to 0. The iterative process is stopped when $|\lambda^{(k+1)}|$ is sufficiently small compared to $||\mathbf{C}||$, where $||\mathbf{C}||$ denotes a matrix norm of matrix $\mathbf{C}$. The proof of convergence of the modified Prony algorithm can be found in Kundu [15].

3.10 Constrained Maximum Likelihood Method

In the IQML or the modified Prony algorithm, the symmetric structure of vector $\mathbf{g}$ as derived in (2.22) has not been utilized. The constrained MLEs, proposed by Kannan and Kundu [16], utilized that symmetric structure of vector $\mathbf{g}$. The problem is same as to minimize $\Psi(\mathbf{g})$ as given in (3.26) with respect to $g_0, g_1, \ldots, g_{2p}$.

   Again differentiating $\Psi(\mathbf{g})$ with respect to $g_0, \ldots, g_{2p}$ and equating them to zero lead to matrix equation of the form

$$\mathbf{C}(\mathbf{x})\mathbf{x}^T = \mathbf{0}, \tag{3.29}$$

here $\mathbf{C}$ is a $(p+1) \times (p+1)$ matrix and $\mathbf{x} = (x_0, \ldots, x_p)^T$ vector. The elements of matrix $\mathbf{C}$ say c_{ij} for $i, j = 0, \ldots, p$, are as follows;

$$
\begin{aligned}
c_{ij} = & \mathbf{Y}^T \mathbf{U}_i (\mathbf{B}^T \mathbf{B})^{-1} \mathbf{U}_j^T \mathbf{Y} + \mathbf{Y}^T \mathbf{U}_j (\mathbf{B}^T \mathbf{B})^{-1} \mathbf{U}_i^T \mathbf{Y} \\
& - \mathbf{Y}^T \mathbf{B} (\mathbf{B}^T \mathbf{B})^{-1} (\mathbf{U}_i^T \mathbf{U}_j + \mathbf{U}_j^T \mathbf{U}_i)(\mathbf{B}^T \mathbf{B})^{-1} \mathbf{B}^T \mathbf{Y}.
\end{aligned}
$$

Here matrix $\mathbf{B}$ is same as defined in (3.19), with g_{p+k} being replaced by g_{p-k}, for $k = 1, \ldots, p$. $\mathbf{U}_1, \ldots, \mathbf{U}_p$ are $n \times (n-2p)$ matrices with entries 0 and 1 only, such that

$$\mathbf{B} = \sum_{j=0}^{p} g_j \mathbf{U}_j.$$

Similar iterative scheme as the modified Prony algorithm has been used to solve for $\widehat{\mathbf{x}} = (\widehat{x}_0, \ldots, \widehat{x}_p)^T$, the solution of (3.29). Once $\widehat{\mathbf{x}}$ is obtained, $\widehat{\mathbf{g}}$ can be easily obtained as follows,

$$\widehat{\mathbf{g}} = (\widehat{x}_0, \ldots, \widehat{x}_{p-1}, \widehat{x}_p, \widehat{x}_{p-1}, \ldots, \widehat{x}_0)^T.$$

From $\widehat{\mathbf{g}}$, the estimates of $\omega_1, \ldots, \omega_p$ can be obtained along the same line as before. The proof of convergence of the algorithm has been established by Kannan and Kundu [16]. The performances of the constrained MLEs are very good as expected.

3.11 Expectation Maximization Algorithm

Expectation Maximization (EM) algorithm, developed by Dempster et al. [17], is a general method for solving the maximum likelihood estimation problem when the data is incomplete. The details on EM algorithm can be found in a book by McLachlan and Krishnan [18]. Although this algorithm has been originally used for incomplete data, sometimes it can be used quite successfully even when the data is complete. It has been used quite effectively to estimate unknown parameters of model (3.1) by Feder and Weinstein [19] under the assumption that the errors are i.i.d. normal random variables.

For a better understanding, we briefly explain the EM algorithm here. Let $\mathbf{Y}$ denote the observed (may be incomplete) data with the probability density function $f_{\mathbf{Y}}(\mathbf{y}; \boldsymbol{\theta})$ indexed by the parameter vector $\boldsymbol{\theta} \in \boldsymbol{\Theta} \subset \mathbb{R}^k$ and let $\mathbf{X}$ denote the complete data vector related to $\mathbf{Y}$ by

$$H(\mathbf{X}) = \mathbf{Y},$$

where $H(\cdot)$ is a many-to-one non-invertible function. Therefore, the density function of $\mathbf{X}$, say $f_{\mathbf{X}}(\mathbf{x}, \boldsymbol{\theta})$, can be written as

$$f_{\mathbf{X}}(\mathbf{x}; \boldsymbol{\theta}) = f_{\mathbf{X}|\mathbf{Y}=\mathbf{y}}(\mathbf{x}; \boldsymbol{\theta}) f_{\mathbf{Y}}(\mathbf{y}; \boldsymbol{\theta}) \quad \forall H(\mathbf{x}) = \mathbf{y}. \tag{3.30}$$

Here $f_{\mathbf{X}|\mathbf{Y}=\mathbf{y}}(\mathbf{x}; \boldsymbol{\theta})$ is the conditional probability density function of $\mathbf{X}$, given $\mathbf{Y} = \mathbf{y}$. From (3.30) after taking the logarithm on both sides, we obtain

$$\ln f_{\mathbf{Y}}(\mathbf{y}; \boldsymbol{\theta}) = \ln f_{\mathbf{X}}(\mathbf{x}; \boldsymbol{\theta}) - \ln f_{\mathbf{X}|\mathbf{Y}=\mathbf{y}}(\mathbf{x}; \boldsymbol{\theta}). \tag{3.31}$$

Taking the conditional expectation, given $\mathbf{Y} = \mathbf{y}$ at the parameter value $\boldsymbol{\theta}'$ on both sides of (3.31), yields

$$\ln f_{\mathbf{Y}}(\mathbf{y}; \boldsymbol{\theta}) = E\{\ln f_{\mathbf{X}}(\mathbf{x}; \boldsymbol{\theta}) | \mathbf{Y} = \mathbf{y}, \boldsymbol{\theta}'\} - E\{\ln f_{\mathbf{X}|\mathbf{Y}=\mathbf{y}}(\mathbf{x}; \boldsymbol{\theta}) | \mathbf{Y} = \mathbf{y}, \boldsymbol{\theta}'\}. \tag{3.32}$$

If we define $L(\boldsymbol{\theta}) = \ln f_{\mathbf{Y}}(\mathbf{y}; \boldsymbol{\theta})$, $U(\boldsymbol{\theta}, \boldsymbol{\theta}') = E\{\ln f_{\mathbf{X}}(\mathbf{x}; \boldsymbol{\theta}) | \mathbf{Y} = \mathbf{y}, \boldsymbol{\theta}'\}$, and $V(\boldsymbol{\theta}, \boldsymbol{\theta}') = E\{\ln f_{\mathbf{X}|\mathbf{Y}=\mathbf{y}}(\mathbf{x}; \boldsymbol{\theta}) | \mathbf{Y} = \mathbf{y}, \boldsymbol{\theta}'\}$, then (3.32) becomes

$$L(\boldsymbol{\theta}) = U(\boldsymbol{\theta}, \boldsymbol{\theta}') - V(\boldsymbol{\theta}, \boldsymbol{\theta}').$$

Here $L(\boldsymbol{\theta})$ is the log-likelihood function of the observed data and that needs to be maximized to obtain MLEs of $\boldsymbol{\theta}$. Since due to Jensen's inequality, see, for example, Chung [20], $V(\boldsymbol{\theta}, \boldsymbol{\theta}') \leq V(\boldsymbol{\theta}', \boldsymbol{\theta}')$, therefore, if,

$$U(\boldsymbol{\theta}, \boldsymbol{\theta}') > U(\boldsymbol{\theta}', \boldsymbol{\theta}'),$$

then

$$L(\boldsymbol{\theta}) > L(\boldsymbol{\theta}'). \tag{3.33}$$

The relation (3.33) forms the basis of the EM algorithm. The algorithm starts with an initial guess and we denote it by $\widehat{\boldsymbol{\theta}}^{(m)}$, the current estimate of $\boldsymbol{\theta}$ after m-iterations. Then $\widehat{\boldsymbol{\theta}}^{(m+1)}$ can be obtained as follows;

$$E \text{ step : Compute } U(\boldsymbol{\theta}, \widehat{\boldsymbol{\theta}}^{(m)})$$
$$M \text{ step : Compute } \widehat{\boldsymbol{\theta}}^{(m)} = \arg \max_{\boldsymbol{\theta}} U(\boldsymbol{\theta}, \widehat{\boldsymbol{\theta}}^{(m)}).$$

Now we show how the EM algorithm can be used to estimate the unknown frequencies and amplitudes of model (3.1) when the errors are i.i.d. normal random variables with mean zero and variance σ^2. Under these assumptions, the log-likelihood function without the constant term takes the following form;

$$l(\boldsymbol{\omega}) = -n \ln \sigma - \sum_{t=1}^{n} \frac{1}{\sigma^2} \left(y(t) - \sum_{k=1}^{p} (A_k \cos(\omega_k t) + B_k \sin(\omega_k t)) \right)^2. \tag{3.34}$$

It is clear that if $\widehat{A}_k$, $\widehat{B}_k$, and $\widehat{\omega}_k$ are the MLEs of A_k, B_k, and ω_k respectively, for $k = 1, \ldots, p$, then the MLE of σ^2 can be obtained as

$$\widehat{\sigma}^2 = \frac{1}{n} \sum_{t=1}^{n} \left(y(t) - \sum_{k=1}^{p} (\widehat{A}_k \cos(\widehat{\omega}_k t) + \widehat{B}_k \sin(\widehat{\omega}_k t)) \right)^2 .$$

It is clear that the MLEs of A_k, B_k, and ω_k for $k = 1, \ldots, p$, can be obtained by minimizing

$$\frac{1}{\sigma^2} \sum_{t=1}^{n} \left(y(t) - \sum_{k=1}^{p} (A_k \cos(\omega_k t) + B_k \sin(\omega_k t)) \right)^2 , \tag{3.35}$$

with respect to the unknown parameters. EM algorithm can be developed to compute the MLEs of A_k, B_k, and ω_k for $k = 1, \ldots, p$, in this case. In developing the EM algorithm, Feder and Weinstein [19] assumed that the noise variance σ^2 is known, and without loss of generality it can be taken as 1.

To implement EM algorithm, re-write the data vector $\mathbf{y}(t)$ as follows:

$$\mathbf{y}(t) = \left(y_1(t), \ldots, y_p(t) \right)^T , \tag{3.36}$$

where

$$y_k(t) = A_k \cos(\omega_k t) + B_k \sin(\omega_k t) + X_k(t).$$

Here, $X_k(t)$ for $k = 1, \ldots, p$ are obtained by arbitrarily decomposing the total noise $X(t)$ into p components, so that

$$\sum_{k=1}^{p} X_k(t) - X(t).$$

Therefore, if $\mathbf{H} = [1, \ldots, 1]$ is a $p \times 1$ vector, then model (3.1) can be written as

$$y(t) = \sum_{k=1}^{p} y_k(t) = \mathbf{H}\mathbf{y}(t).$$

If we choose $X_1(t), \ldots, X_p(t)$ to be independent normal random variables with mean zero and variance $\beta_1, \ldots, \beta_p$, respectively, then

$$\sum_{i=1}^{p} \beta_k = 1, \quad \beta_k \geq 0.$$

With the above notation, the EM algorithm takes the following form. If $\widehat{A}_k^{(m)}$, $\widehat{B}_k^{(m)}$, and $\widehat{\omega}_k^{(m)}$ denote the estimates of A_k, B_k, and ω_k, respectively, after m-iterations, then

E-Step:

$$\widehat{y}_k^{(m)}(t) = \widehat{A}_k^{(m)} \cos(\widehat{\omega}_k^{(m)}t) + \widehat{B}_k^{(m)} \sin(\widehat{\omega}_k^{(m)}t)$$
$$+\beta_k \left[y(t) - \sum_{k=1}^{p} \widehat{A}_k^{(m)} \cos(\widehat{\omega}_k^{(m)}t) - \widehat{B}_k^{(m)} \sin(\widehat{\omega}_k^{(m)}t) \right]. \quad (3.37)$$

M-Step:

$$(\widehat{A}_k^{(m+1)}, \widehat{B}_k^{(m+1)}, \widehat{\omega}_k^{(m+1)}) = \arg \min_{A,B,\omega} \sum_{t=1}^{n} \left(\widehat{y}_k^{(m)}(t) - A \cos(\omega t) - B \sin(\omega t) \right)^2. \quad (3.38)$$

It is interesting to observe that $\widehat{A}_k^{(m+1)}$, $\widehat{B}_k^{(m+1)}$, and $\widehat{\omega}_k^{(m+1)}$ are the MLEs of A_k, B_k, and ω_k, respectively, based on $\widehat{y}_k^{(m)}(t), t = 1, \ldots, n$. The most important feature of this algorithm is that it decomposes the complicated optimization problem into p separate simple 1-D optimization problem.

Feder and Weinstein [19] did not mention how to choose $\beta_1, \ldots, \beta_p$ and how the EM algorithm can be used when the error variance σ^2 is unknown. The choice of $\beta_1, \ldots, \beta_p$ plays an important role in the performance of the EM algorithm. One choice might be to take $\beta_1 = \cdots = \beta_p$, alternatively dynamical choice of $\beta_1, \ldots, \beta_p$ might provide better results.

We propose the following EM algorithm when σ^2 is unknown. Suppose $\widehat{A}_k^{(m)}$, $\widehat{B}_k^{(m)}$, $\widehat{\omega}_k^{(m)}$, and $\widehat{\sigma}^{2(m)}$ are the estimates of A_k, B_k, ω_k, and σ^2, respectively, at the m-step of the EM algorithm. They may be obtained from the periodogram estimates. Choose $\beta_k^{(m)}$ as

$$\beta_k^{(m)} = \frac{\widehat{\sigma}^{2(m)}}{p}; \quad k = 1, \ldots, p.$$

In E-step of (3.37) replace β_k by $\beta_k^{(m)}$, and in the M-step after computing (3.38) also obtain

$$\widehat{\sigma}^{2(m+1)} = \frac{1}{n} \sum_{t=1}^{n} \left(y(t) - \sum_{k=1}^{p} \widehat{A}_k^{(m+1)} \cos(\widehat{\omega}_k^{(m+1)}t) + \widehat{B}_k^{(m+1)} \sin(\widehat{\omega}_k^{(m+1)}t) \right)^2.$$

The iteration continues unless convergence criterion is met. The proof of convergence or the properties of the estimators have not yet been established. Further work is needed along that direction.

3.12 Sequential Estimators

One major drawback of different estimators discussed so far is in their computational complexity. Sometimes to reduce the computational complexity, efficiency has been sacrificed. Prasad et al. [21] suggested sequential estimators, where computational complexity has been reduced and at the same time the efficiency of the estimators has not been sacrificed.

Prasad et al. [21] considered model (3.1) under error Assumptions 3.1 and 3.2. The sequential method is basically a modification of the approximate least squares method as described in Sect. 3.1. Using the same notation as in (3.5), the method can be described as follows:

Algorithm 3.3

- Step 1: Compute $\widehat{\omega}_1$, which can be obtained by minimizing $R_1(\omega)$, with respect to ω, where

$$R_1(\omega) = \mathbf{Y}^T (\mathbf{I} - \mathbf{P}_{\mathbf{Z}(\omega)}) \mathbf{Y}. \tag{3.39}$$

Here $\mathbf{Z}(\omega)$ is same as defined in Sect. 3.1, and $\mathbf{P}_{\mathbf{Z}(\omega)}$ is the projection matrix on the column space of $\mathbf{Z}(\omega)$.
- Step 2: Construct the following vector

$$\mathbf{Y}^{(1)} = \mathbf{Y} - \mathbf{Z}(\widehat{\omega}_1)\widehat{\alpha}_1, \tag{3.40}$$

where

$$\widehat{\alpha}_1 = \left[\mathbf{Z}^T(\widehat{\omega}_1)\mathbf{Z}(\widehat{\omega}_1) \right]^{-1} \mathbf{Z}^T(\widehat{\omega}_1)\mathbf{Y}.$$

- Step 3: Compute $\widehat{\omega}_2$, which can be obtained by minimizing $R_2(\omega)$, with respect to ω, where $R_2(\omega)$ is obtained by replacing $\mathbf{Y}$ with $\mathbf{Y}^{(1)}$ in (3.39).
- Step 4: The process continues up to p-steps.

The main advantage of the proposed algorithm is that it significantly reduces the computational burden. The minimization of $R_k(\omega)$ for each k is a 1-D optimization process, and it can be performed quite easily. It has been shown by the authors that the estimators obtained by this sequential procedure are strongly consistent and they have the same rate of convergence as the LSEs. Moreover, if the process continues even after p-steps, it has been shown that the estimators of A and B converge to zero a.s.

3.13 Quinn and Fernandes Method

Quinn and Fernandes [22] proposed the following method by exploiting the fact that there is a second-order filter which annihilates a sinusoid at a given frequency. First consider model (3.1) with $p = 1$. It follows from Prony's equation that the model satisfies the following equation;

$$y(t) - 2\cos(\omega)y(t-1) + y(t-2) = X(t) - 2\cos(\omega)X(t-1) + X(t-2).$$

$$(3.41)$$

Therefore, $\{y(t)\}$ forms an autoregressive moving average, namely ARMA(2,2) process. It may be noted that the process does not have a stationary or invertible solution. As expected, the above process does not depend on the linear parameters A and B, but only depends on the non-linear frequency ω. It is clear from (3.41) that using the above ARMA(2,2) structure of $\{y(t)\}$, it is possible to obtain the estimate of ω.

Re-write (3.41) as follows;

$$y(t) - \beta y(t-1) + y(t-2) = X(t) - \alpha X(t-1) + X(t-2), \qquad (3.42)$$

and the problem is to estimate the unknown parameter with the constraint $\alpha = \beta$, based on the observation $\{y(t)\}$ for $t = 1, \ldots, n$. It is important to note that if the standard ARMA-based technique is used to estimate the unknown parameter, it can only produce estimator which has the asymptotic variance of $O(n^{-1})$ the order n^{-1}. On the other hand, it is known that the LSE of ω has the asymptotic variance of $O(n^{-3})$ the order n^{-3}. Therefore, some 'non-standard' ARMA-based technique needs to be used to obtain an efficient frequency estimator.

If α is known, and $X(1), \ldots, X(n)$ are i.i.d. normal random variables, then the MLE of β can be obtained by minimizing

$$Q(\beta) = \sum_{t=1}^{n} (\xi(t) - \beta\xi(t-1) + \xi(t-2))^2, \qquad (3.43)$$

with respect to β, where $\xi(t) = 0$ for $t < 1$, and for $t \geq 1$,

$$\xi(t) = y(t) + \alpha\xi(t-1) - \xi(t-2).$$

The value of β, which minimizes (3.43), can easily be obtained as

$$\alpha + \frac{\sum_{t=1}^{n} y(t)\xi(t-1)}{\sum_{t=1}^{n} \xi^2(t-1)}. \qquad (3.44)$$

Therefore, one way can be to put the new value of α in (3.44) and then re-estimate β. This basic idea has been used by Quinn and Fernandes [22] with a proper acceleration factor, which ensures the convergence of the iterative procedure also. The algorithm is as follows;

Algorithm 3.4

- Step 1: Put $\alpha^{(1)} = 2\cos(\widehat{\omega}^{(1)})$, where $\widehat{\omega}^{(1)}$ is an initial estimator of ω.
- Step 2: For $j \geq 1$, compute

$$\xi(t) = y(t) + \alpha^{(j)}\xi(t-1) - \xi(t-2); \quad t = 1, \ldots, n,$$

where $\xi(t) = 0$ for $t < 1$.

- Step 3: Obtain

$$\beta^{(j)} = \alpha^{(j)} + 2\frac{\sum_{t=1}^{n} y(t)\xi(t-1)}{\sum_{t=1}^{n} \xi^2(t-1)}.$$

- Step 4: If $|\alpha^{(j)} - \beta^{(j)}|$ is small then stop the iteration procedure, and obtain estimate of ω as $\widehat{\omega} = \cos^{-1}(\beta^{(j)}/2)$. Otherwise obtain $\alpha^{(j+1)} = \beta^{(j)}$ and go back to Step 2.

In the same paper, the authors extended the algorithm for general model (3.1) also based on the observation that a certain difference operator annihilates all the sinusoidal components. If $y(1), \ldots, y(n)$ are obtained from model (3.1), then from Prony's equations it again follows that there exists $\alpha_0, \ldots, \alpha_{2p}$, so that

$$\sum_{j=0}^{2p} \alpha_j y(t-j) = \sum_{j=0}^{2p} \alpha_j X(t-j), \tag{3.45}$$

where

$$\sum_{j=0}^{2p} \alpha_j z^j = \prod_{j=1}^{p} (1 - 2\cos(\omega_j)z + z^2). \tag{3.46}$$

It is clear from (3.46) that $\alpha_0 = \alpha_{2p} = 1$, and $\alpha_{2p-j} = \alpha_j$ for $j = 0, \ldots, p-1$. Therefore, in this case $y(1), \ldots, y(n)$ form an ARMA($2p, 2p$) process, and all the zeros of the corresponding auxiliary polynomial are on the unit circle. It can also be observed from Prony's equations that no other polynomial of order less than $2p$ has this property.

Following exactly the same reasoning as before, Quinn and Fernandes [22] suggested the following algorithm for multiple sinusoidal model;

Algorithm 3.5

- Step 1: If $\widetilde{\omega}_1, \ldots, \widetilde{\omega}_p$ are initial estimators of $\omega_1, \ldots, \omega_p$, compute $\alpha_1, \ldots, \alpha_p$ from

$$\sum_{j=0}^{2p} \alpha_j z^j = \prod_{j=1}^{p} (1 - 2z\cos(\widetilde{\omega}_j) + z^2).$$

- Step 2: Compute for $t = 1, \ldots, n$;

$$\xi(t) = y(t) - \sum_{j=1}^{2p} \alpha_j \xi(t-j)$$

and for $j = 1, \ldots, p - 1$ compute the $p \times 1$ vector

$$\eta(t - 1) = \left[\widetilde{\xi}(t - 1), \ldots, \widetilde{\xi}(t - p + 1), \quad \xi(t - p) \right]^T,$$

where

$$\widetilde{\xi}(t - j) = \xi(t - j) + \xi(t - 2p + j),$$

and $\xi(t) = 0$, for $t < 1$.

- Step 3: Let $\alpha = \left[\alpha_1, \ldots, \alpha_p \right]^T$, compute $\beta = \left[\beta_1, \ldots, \beta_p \right]^T$, where

$$\beta = \alpha - 2 \left\{ \sum_{t=1}^{n} \eta(t - 1)\eta(t - 1)^T \right\}^{-1} \sum_{t=1}^{n} y(t)\eta(t - 1).$$

- Step 4: If $\max_j |\beta_j - \alpha_j|$ is small stop the iteration and obtain estimates of $\omega_1, \ldots, \omega_p$. Otherwise set $\alpha = \beta$ and go to Step 2.

This algorithm works very well for small p. For large p, the performance is not very satisfactory, as it involves a $2p \times 2p$ matrix inversion, which is quite ill conditioned. Due to this reason, the elements of the vectors α and β obtained by this algorithm can be quite large.

Quinn and Fernandes [22] suggested the following modified algorithm that works very well even for large p also.

Algorithm 3.6

- Step 1: If $\widetilde{\omega}_1, \ldots, \widetilde{\omega}_p$ are initial estimators of $\omega_1, \ldots, \omega_p$, respectively, compute $\theta_k = 2\cos(\widetilde{\omega}_k)$, for $k = 1, \ldots, p$.
- Step 2: Compute for $t = 1, \ldots, n, j = 1, \ldots, p, \zeta_j(t)$, where $\zeta_j(-1) = \zeta_j(-2) = 0$, and they satisfy

$$\zeta_j(t) - \theta_j \zeta_j(t - 1) + \zeta_j(t - 2) = y(t).$$

- Step 3: Compute $\theta = \left[\theta_1, \ldots, \theta_p \right]^T, \zeta(t) = \left[\zeta_1(t), \ldots, \zeta_p(t) \right]^T$, and

$$\psi = \theta + 2 \left\{ \sum_{t=1}^{n} \zeta(t - 1)\zeta(t - 1)^T \right\}^{-1} \sum_{t=1}^{n} y(t)\zeta(t - 1).$$

- Step 4: If $|\psi - \theta|$ is small, stop the iteration and obtain the estimates of the $\omega_1, \cdots, \omega_p$. Otherwise set $\theta = \psi$ and go to Step 2.

Comments: Note that in both the above algorithms which have been proposed by Quinn and Fernandes [22] for $p > 1$ involve the computation of a $2p \times 2p$ matrix inversion. Therefore, if p is very large then it becomes a computationally challenging problem.

3.14 Amplified Harmonics Method

The estimation of frequencies of model (3.1) based on amplified harmonics was proposed by Truong-Van [23]. The main idea of Truong-Van [23] is to construct a process which enhances the amplitude of a particular frequency quasi-linearly with t, whereas the amplitudes of the other frequencies remain constant in time. The method is as follows. For each frequency ω_k, and any estimate $\omega_k^{(0)}$ near ω_k, define the process $\xi(t)$ as the solution of the following linear equation;

$$\xi(t) - 2\alpha_k^{(0)}\xi(t-1) + \xi(t-2) = y(t); \quad t \geq 1, \tag{3.47}$$

with the initial conditions $\xi(0) = \xi(-1) = 0$ and $\alpha_k^{(0)} = \cos(\omega_k^{(0)})$. Using the results of Ahtola and Tiao [24], it can be shown that

$$\xi(t) = \sum_{j=0}^{t-1} v(j; \omega_k^{(0)})y(t-k); \quad t \geq 1, \tag{3.48}$$

where

$$v(j; \omega) = \frac{\sin(\omega(j+1))}{\sin(\omega)}.$$

Note that the process $\xi(t)$ depends on $\omega_k^{(0)}$, but we do not make it explicit. If it is needed, we denote it by $\xi(t; \omega_k^{(0)})$. It has been shown by the author that the process $\xi(t)$ acts like an amplifier of the frequency ω_k. It has been further shown that such an amplifier exists and proposed an estimator $\widehat{\omega}_k$ of ω_k by observing the fact that $y(t)$ and $\xi(t-1; \widehat{\omega}_k)$ are orthogonal to each other, that is,

$$\sum_{t=2}^{n} \xi(t-1; \widehat{\omega}_k)y(t) = 0. \tag{3.49}$$

Truong-Van [23] proposed two different algorithms mainly to solve (3.49). The first algorithm (Algorithm 3.7) has been proposed when the initial guess values are very close to the true values and the algorithm is based on Newton's method to solve (3.49). The second algorithm (Algorithm 3.8) has been proposed when the initial guess values are not very close to the true values, and this algorithm also tries to find a solution of the non-linear equation (3.49) using least squares approach.

Algorithm 3.7

- Step 1: Find an initial estimator $\omega_k^{(0)}$ of ω_k.
- Step 2: Compute

$$\omega_k^{(1)} = \omega_k^{(0)} - F(\omega_k^{(0)})F'(\omega_k^{(0)})^{-1},$$

where

$$F(\omega) = \sum_{t=2}^{n} \xi(t - 1; \omega) y(t), \quad \text{and} \quad F'(\omega) = \sum_{t=2}^{n} \frac{d}{d\omega} \xi(t - 1; \omega) y(t).$$

- Step 3: If $\omega_k^{(0)}$ and $\omega_k^{(1)}$ are close to each other, stop the iteration, otherwise replace $\omega_k^{(0)}$ by $\omega_k^{(1)}$ and continue the process.

Algorithm 3.8

- Step 1: Find an initial estimator $\omega_k^{(0)}$ of ω_k and compute $\alpha_k^{(0)} = \cos(\omega_k^{(0)})$.
- Step 2: Compute

$$\alpha_k^{(1)} = \alpha_k^{(0)} + \left(2 \sum_{t=2}^{n} \xi^2(t - 1; \omega_k^{(0)}) \right)^{-1} F(\omega_k^{(0)}).$$

- Step 3: If $\alpha_k^{(0)}$ and $\alpha_k^{(1)}$ are close to each other, stop the iteration, otherwise replace $\alpha_k^{(0)}$ by $\alpha_k^{(1)}$ and continue the process. From the estimate of α_k, the estimate of ω_k can be easily obtained.

3.15 Weighted Least Squares Estimators

Weighted least squares estimators (WLSEs) are proposed by Irizarry [25]. The main idea is to produce asymptotically unbiased estimators, which may have lower variances than the LSEs depending on the weight function. Irizarry [25] considered model (3.1), and WLSEs of the unknown parameters can be obtained by minimizing

$$S(\boldsymbol{\omega}, \boldsymbol{\theta}) = \sum_{t=1}^{n} w \left(\frac{t}{n} \right) \left(y(t) - \sum_{k=1}^{p} \{ A_k \cos(\omega_k t) + B_k \sin(\omega_k t) \} \right)^2, \qquad (3.50)$$

with respect to $\boldsymbol{\omega} = (\omega_1, \ldots, \omega_p), \boldsymbol{\theta} = (A_1, \ldots, A_p, B_1, \ldots, B_p)$. The weight function $w(s)$ is non-negative, of bounded variation, has support $[0, 1]$. Moreover, it is such that $W_0 > 0$ and $W_1^2 - W_0 W_2 \neq 0$, where

$$W_n = \int_0^1 s^n w(s) ds.$$

It is assumed that the weight function is known a priori. In this case, it can be seen along the same line as the LSEs that if we denote $\widehat{\omega}_1, \ldots, \widehat{\omega}_k, \widehat{A}_1, \ldots, \widehat{A}_k, \widehat{B}_1, \ldots, \widehat{B}_k$

as the WLSEs of $\omega_1, \ldots, \omega_k, A_1, \ldots, A_k, B_1, \ldots, B_k$, respectively, then they can be obtained as follows. First obtain $\widehat{\omega}_1, \ldots, \widehat{\omega}_k$, which maximize $Q(\boldsymbol{\omega})$, with respect to $\omega_1, \ldots, \omega_k$, where

$$Q(\boldsymbol{\omega}) = \sum_{k=1}^{p} \left| \frac{1}{n} \sum_{t=1}^{n} w\left(\frac{t}{n}\right) y(t) e^{it\omega_k} \right|^2, \tag{3.51}$$

and then $\widehat{A}_k$ and $\widehat{B}_k$ are obtained as

$$\widehat{A}_k = \frac{2\sum_{t=1}^{n} w\left(\frac{t}{n}\right) y(t) \cos(\widehat{\omega}_k t)}{\sum_{t=1}^{n} w\left(\frac{t}{n}\right)} \quad \text{and} \quad \widehat{B}_k = \frac{2\sum_{t=1}^{n} w\left(\frac{t}{n}\right) y(t) \sin(\widehat{\omega}_k t)}{\sum_{t=1}^{n} w\left(\frac{t}{n}\right)},$$

for $k = 1, \ldots, p$. Irizarry [25] proved that WLSEs are strongly consistent estimators of the corresponding parameters, and they are asymptotically normally distributed under fairly general assumptions on the weight function and on the error random variables. The explicit expression of the variance covariance matrix is also provided, which is as expected depends on the weight function. It appears that with the proper choice of the weight function, the asymptotic variances of the WLSEs can be made smaller than the corresponding asymptotic variances of the LSEs, although it has not been explored. Moreover, it has not been indicated how to maximize $Q(\omega)$ as defined in (3.51), with respect to the unknown parameter vector ω. It is a multi-dimensional optimization problem, and if p is large, it is a difficult problem to solve. It might be possible to use the sequential estimation procedure as suggested by Prasad et al. [21], see Sect. 3.12, in this case also. More work is needed in that direction.

3.16 Nandi and Kundu Algorithm

Nandi and Kundu [26] proposed a computationally efficient algorithm for estimating the parameters of sinusoidal signals in the presence of additive stationary noise, that is, under error Assumption 3.2. The key features of the proposed algorithm are (i) the estimators are strongly consistent and they are asymptotically equivalent to the LSEs, (ii) the algorithm converges in three steps starting from the initial frequency estimators as the PEs over Fourier frequencies, and (iii) the algorithm does not use the whole sample at each step. In the first two steps, it uses only some fractions of the whole sample, only in the third step it uses the whole sample. For notational simplicity, we describe the algorithm when $p = 1$, and note that for general p, the sequential procedure of Prasad et al. [21] can be easily used.

If at the jth stage the estimator of ω is denoted by $\omega^{(j)}$, then $\omega^{(j+1)}$ is calculated as

$$\omega^{(j+1)} = \omega^{(j)} + \frac{12}{n_j} \, \text{Im}\left[\frac{P(j)}{Q(j)}\right], \tag{3.52}$$

where

$$P(j) = \sum_{t=1}^{n_j} y(t) \left(t - \frac{n_j}{2} \right) e^{-i\omega^{(j)}t}, \tag{3.53}$$

$$Q(j) = \sum_{t=1}^{n_j} y(t) e^{-i\omega^{(j)}t}, \tag{3.54}$$

and n_j denotes the sample size used at the jth iteration. Suppose $\omega^{(0)}$ denotes the periodogram estimator of ω, then the algorithm takes the following form.

Algorithm 3.9

1. Step 1: Compute $\omega^{(1)}$ from $\omega^{(0)}$ using (3.52) with $n_1 = n^{0.8}$.
2. Step 2: Compute $\omega^{(2)}$ from $\omega^{(1)}$ using (3.52) with $n_2 = n^{0.9}$.
3. Step 3: Compute $\omega^{(3)}$ from $\omega^{(2)}$ using (3.52) with $n_3 = n$.

It should be mentioned that the fraction 0.8 or 0.9, which has been used in Step 2 or Step 3, respectively, is not unique, and several other choices are also possible, see Nandi and Kundu [26] for details. Moreover, it has been shown by the authors that asymptotic properties of $\omega^{(3)}$ are same as the corresponding LSE.

3.17 Super Efficient Estimator

Kundu et al. [27] recently proposed a modified Newton–Raphson method to obtain super efficient estimators of the frequencies of model (3.1) in the presence of stationary noise $\{X(t)\}$. It is observed that if the algorithm starts with an initial estimator with a convergence rate $O_p(1/n)$, and uses the Newton–Raphson algorithm with proper step factor modification, then it produces super efficient frequency estimator, in the sense that its asymptotic variance is lower than the asymptotic variance of the corresponding LSE. It is indeed a very counterintuitive result because it is well known that the usual Newton–Raphson method cannot be used to compute the LSE, whereas with proper step-factor modification, it can produce super efficient frequency estimator.

If we denote

$$S(\omega) = \mathbf{Y}^T \mathbf{Z} (\mathbf{Z}^T \mathbf{Z})^{-1} \mathbf{Z}^T \mathbf{Y}, \tag{3.55}$$

where $\mathbf{Y}$ and $\mathbf{Z}$ are same as defined in (3.5), then the LSE of ω can be obtained by maximizing $S(\omega)$ with respect to ω. The maximization of $S(\omega)$ using Newton–Raphson algorithm can be performed as follows:

$$\omega^{(j+1)} = \omega^{(j)} - \frac{S'(\omega^{(j)})}{S''(\omega^{(j)})}, \tag{3.56}$$

here $\omega^{(j)}$ is same as defined before, that is, the estimate of ω at the jth stage, moreover, $S'(\omega^{(j)})$ and $S''(\omega^{(j)})$ denote the first derivative and second derivative, respectively, of $S(\omega)$ evaluated at $\omega^{(j)}$. The standard Newton–Raphson algorithm is modified with a smaller correction factor as follows:

$$\omega^{(j+1)} = \omega^{(j)} - \frac{1}{4} \times \frac{S'(\omega^{(j)})}{S''(\omega^{(j)})}. \tag{3.57}$$

Suppose $\omega^{(0)}$ denotes the PE of ω, then the algorithm can be described as follows:

Algorithm 3.10

(1) Take $n_1 = n^{6/7}$, and calculate

$$\omega^{(1)} = \omega^{(0)} - \frac{1}{4} \times \frac{S'_{n_1}(\omega^{(0)})}{S''_{n_1}(\omega^{(0)})},$$

where $S'_{n_1}(\omega^{(0)})$ and $S''_{n_1}(\omega^{(0)})$ are same as $S'(\omega^{(0)})$ and $S''(\omega^{(0)})$, respectively, computed using a subsample of size n_1.
(2) With $n_j = n$, repeat

$$\omega^{(j+1)} = \omega^{(j)} - \frac{1}{4} \times \frac{S'_{n_j}(\omega^{(j)})}{S''_{n_j}(\omega^{(j)})}, \quad j = 1, 2, \ldots,$$

until a suitable stopping criterion is satisfied.

It is observed that any n_1 consecutive data points can be used at step (1) to start the algorithm, and it is observed in the simulation study that the choice of the initial subsamples does not have any visible effect on the final estimator. Moreover, the factor $6/7$ in the exponent at step (1) is not unique, and there are several other ways the algorithm can be initiated. It is observed in extensive simulation studies by Kundu et al. [27] that the iteration converges very quickly and it produces frequency estimator which has lower variances than the corresponding LSE.

3.18 Conclusions

In this section, we discussed different estimation procedures for estimating the frequencies of the sum of sinusoidal models. It should be mentioned that although we have discussed 17 different methods, the list is no where near complete. The main aim is to provide an idea how the same problem due to its complicated nature has been attempted by different methods to get some satisfactory answers. Moreover, it is also observed that none of these methods work uniformly well for all values of the model parameters. It is observed that finding the efficient estimators is a numerically challenging problem. Due to this reason several suboptimal solutions have been suggested in the literature which do not have the same rate of convergence as the efficient estimators. The detailed theoretical properties of the different estimators are provided in the next chapter.

References

1. Rice, J. A., & Rosenblatt, M. (1988). On frequency estimation. *Biometrika,75*, 477–484.
2. Bai, Z. D., Chen, X. R., Krishnaiah, P. R., & Zhao, L. C. (1987). *Asymptotic properties of EVLP estimators for superimposed exponential signals in noise*. Technical report 87-19, CMA, University of Pittsburgh.
3. Bai, Z. D., Rao, C. R., Chow, M., & Kundu, D. (2003). An efficient algorithm for estimating the parameters of superimposed exponential signals. *Journal of Statistical, Planning and Inference,110*, 23–34.
4. Kumaresan, R. (1982). *Estimating the parameters of exponential signals*. Ph.D. thesis, U. Rhode Island.
5. Tufts, D. W., & Kumaresan, R. (1982). Estimation of frequencies of multiple sinusoids: Making linear prediction perform like maximum likelihood. *Proceedings of the IEEE, 70*, 975–989.
6. Rao, C. R. (1988). Some results in signal detection. In S. S. Gupta & J. O. Berger (Eds.) *Decision theory and related topics, IV* (Vol. 2, pp. 319–332). New York: Springer.
7. Kundu, D., & Mitra, A. (1995). Consistent method of estimating the superimposed exponential signals. *Scandinavian Journal of Statistics, 22*, 73–82.
8. Kundu, D., & Mitra, A. (1997). Consistent methods of estimating sinusoidal frequencies; a non iterative approach. *Journal of Statistical Computation and Simulation, 58*, 171–194.
9. Roy, R. H. (1987). *ESPRIT-estimation of signal parameters via rotational invariance technique*. Ph.D. thesis, Stanford University
10. Roy, R. H., & Kailath, T. (1989). ESPRIT-Estimation of Signal Parameters via Rotational Invariance Technique. *IEEE Transactions on ASSP, 43*, 984–995.
11. Pillai, S. U. (1989). *Array processing*. New York: Springer.
12. Quinn, B. G. (1994). Estimating frequency by interpolation using Fourier coefficients. *IEEE Transactions on Signal process, 42*, 1264–1268.
13. Bresler, Y., & Macovski, A. (1986). Exact maximum likelihood parameter estimation of superimposed exponential signals in noise. *IEEE Transactions on ASSP, 34*, 1081–1089.
14. Kundu, D. (1993). Estimating the parameters of undamped exponential signals. *Technometrics, 35*, 215–218.
15. Kundu, D. (1994). Estimating the parameters of complex valued exponential signals. *Computational Statistics and Data Analysis, 18*, 525–534.
16. Kannan, N., & Kundu, D. (1994). On modified EVLP and ML methods for estimating superimposed exponential signals. *Signal Processing, 39*, 223–233.
17. Dempster, A. P., Laird, N. M., & Rubin, D. B. (1977). Maximum likelihood from incomplete data via EM algorithm. *Journal of the Royal Statistical Society Series B, 39*, 1–38.
18. McLachlan, G. J., & Krishnan, T. (2008). *The EM algorithm and extensions*. Hoboken, New Jersey: Wiley.
19. Feder, M., & Weinstein, E. (1988). Parameter estimation of superimposed signals using the EM algorithm. *IEEE Transactions on ASSP, 36*, 477–489.
20. Chung, K. L. (1974). *A course in probability theory*. New York: Academic Press.
21. Prasad, A., Kundu, D., & Mitra, A. (2008). Sequential estimation of the sum of sinusoidal model parameters. *Journal of Statistical, Planning and Inference, 138*, 1297–1313.
22. Quinn, B. G., & Fernandes, J. M. (1991). A fast efficient technique for the estimation of frequency. *Biometrika, 78*, 489–497.

23. Truong-Van, B. (1990). A new approach to frequency analysis with amplified harmonics. *Journal of the Royal Statistical Society Series B, 52,* 203–221.
24. Ahtola, J., & Tiao, G. C. (1987). Distributions of least squares estimators of autoregressive parameters for a process with complex roots on the unit circle. *Journal of Time Series Analysis, 8,* 1–14.
25. Irizarry, R. A. (2002). Weighted estimation of harmonic components in a musical sound signal. *Journal of Time Series Analysis, 23,* 29–48.
26. Nandi, S., & Kundu, D. (2006). A fast and efficient algorithm for estimating the parameters of sum of sinusoidal model. *Sankhya, 68,* 283–306.
27. Kundu, D., Bai, Z. D., Nandi, S., & Bai, L. (2011). Super efficient frequency eEstimation". *Journal of Statistical, Planning and Inference, 141*(8), 2576–2588.

Chapter 4
Asymptotic Properties

4.1 Introduction

In this chapter, we discuss asymptotic properties of some of the estimators described in Chap. 3. Asymptotic results or results based on large samples deal with properties of estimators under the assumption that the sample size increases indefinitely. The statistical models, observed in signal processing literature, are mostly very complicated non-linear models. Even a single component sinusoidal component model is highly non-linear in its frequency parameter. Due to that many statistical concepts for small samples cannot be applied in case of sinusoidal model. The added problem is that under different error assumptions, this model is mean non-stationary. Therefore, it is not possible to obtain any finite sample property of the LSE or any other estimators, discussed in the previous chapter. All the results have to be asymptotic. The most intuitive estimator is the LSE and the most popular one is the ALSE. These two estimators are asymptotically equivalent and we discuss their equivalence in Sect. 4.4.3. The sinusoidal model is a non-linear regression model, but it does not satisfy, see Kundu [1], the standard sufficient conditions of Jennrich [2] or Wu [3] for the LSE to be consistent. Jennrich [2] first proved the existence of the LSE in a non-linear regression model of the form $y(t) = f_t(\theta^0) + \varepsilon(t), t = 1, \dots$. The almost sure convergence of the LSE of the unknown parameter θ was shown under the following assumption: Define $F_n(\theta_1, \theta_2) = \sum_{t=1}^{n} (f_t(\theta_1) - f_t(\theta_2))^2 / n$, then $F_n(\theta_1, \theta_2)$ converges uniformly to a continuous function $F(\theta_1, \theta_2)$ and $F(\theta_1, \theta_2) \neq 0$ if and only if $\theta_1 = \theta_2$. Consider a single-component sinusoidal model and assume that $A^0 = 1$ and $B^0 = 0$ in (4.1). Suppose the model satisfies Assumption 3.1 and ω^0 is an interior point of $(0, \pi)$. In this simple situation, $F_n(\theta_1, \theta_2)$ does not converge uniformly to a continuous function. Wu [3] gave some sufficient condition under which the LSE of θ^0 is strongly consistent when the growth rate requirement of $F_n(\theta_1, \theta_2)$ is replaced by a Lipschitz-type condition on the sequence $\{f_t\}$. In addition, the sinusoidal model does not satisfy Wu's Lipschitz-type condition also.

D. Kundu and S. Nandi, *Statistical Signal Processing*, SpringerBriefs in Statistics,   45
DOI: 10.1007/978-81-322-0628-6_4, © The Author(s) 2012

Whittle [4] first obtained some of the theoretical results. Recent results are by Hannan [5, 6], Walker [7], Rice and Rosenblatt [8], Quinn and Fernandes [9], Kundu [1, 10], Quinn [11], Kundu and Mitra [12], Irizarry [13], Nandi et al. [14], Prasad et al. [15], and Kundu et al. [16]. Walker [7] considered the sinusoidal model with one component and obtained the asymptotic properties of the PEs under the assumption that the errors are i.i.d. with mean zero and finite variance. The result has been extended by Hannan [5] when the errors are from a weakly stationary process and by Hannan [6] when the errors are from a strictly stationary random process with continuous spectrum. Some of the computational issues have been discussed in Rice and Rosenblatt [8]. The estimation procedure, proposed by Quinn and Fernandes [9], is based on fitting ARMA (2,2) models and the estimator is strongly consistent and efficient. Kundu and Mitra [12] considered the model when errors are i.i.d. and proved directly the consistency and asymptotic normality of the LSEs. The result was generalized by Kundu [10] when the errors are from a stationary linear process. The weighted LSEs are proposed by Irizarry [13] and extended the asymptotic results of the LSEs to the weighted one. Nandi et al. [14] prove the strong consistency of the LSEs when the errors are i.i.d. with mean zero, but may not have finite variance. They also obtain the asymptotic distribution of the LSEs when the error distribution is symmetric stable. It is well known that the Newton–Raphson method does not work well in case of sinusoidal frequency model. Kundu et al. [16] propose a modification of the Newton–Raphson method with a smaller step factor such that the resulting estimator has the same rate of convergence as the LSEs. Additionally, the asymptotic variances of the proposed estimators are less than those of the LSEs. Therefore, the estimators are named as the superefficient estimators.

4.2 Sinusoidal Model with One Component

We first discuss the asymptotic results of the estimators of the parameters of the sinusoidal model with one component. This is just to keep the mathematical expression simple. We talk about the model with p components at the end of the chapter. The model is now

$$y(t) = A^0 \cos(\omega^0 t) + B^0 \sin(\omega^0 t) + X(t), \quad t = 1, \ldots, n. \tag{4.1}$$

In this section, we explicitly write A^0, B^0, and ω^0 as the true values of the unknown parameters A, B, and ω, respectively. Write $\theta = (A, B, \omega)$ and let θ^0 be the true value of θ and let $\widehat{\theta}$ and $\widetilde{\theta}$ be the LSE and ALSE of θ, respectively; $\{X(t)\}$ is a sequence of error random variables. To ensure the presence of the frequency component and to make sure that $y(t)$ is not pure noise, assume that A^0 and B^0 are not simultaneously equal to zero. For technical reason, take $\omega \in (0, \pi)$. At this moment, we do not explicitly mention the complete error structure. We assume in this chapter that the number of signal components (distinct frequency), p, is known in advance. The problem of estimation of p is considered in Chap. 5.

In the following, we discuss the consistency and asymptotic distribution of the LSE and ALSE of θ under different error assumptions. Apart from Assumptions 3.1 and 3.2, the following two assumptions regarding the structure of the error process are required.

Assumption 4.1 $\{X(t)\}$ is a sequence of i.i.d. random variables with mean zero and $E|X(t)|^{1+\delta} < \infty$ for some $0 < \delta \le 1$.

Assumption 4.2 $\{X(t)\}$ is a sequence of i.i.d. random variables distributed as $S\alpha S(\sigma)$.

Note 4.1

(a) Assumption 3.1 is a special case of Assumption 4.1 as both are same with $\delta = 1$.
(b) If $\{X(t)\}$ satisfies Assumption 4.2, Assumption 4.1 is also true with $1 + \delta < \alpha \le 2$. Therefore, from now on, we take $1 + \delta < \alpha \le 2$.
(c) Assumption 3.2 is a standard assumption for a stationary linear process, any finite dimensional stationary AR, MA, or ARMA process can be represented as a linear process with absolute summable coefficients. A process is called a stationary linear process if it satisfies Assumption 3.2.

4.3 Strong Consistency of LSE and ALSE of θ

We recall that $Q(\theta)$ is the residual sum of squares, defined in (3.3) and $I(\omega)$ is the periodogram function, defined in (1.5).

Theorem 4.1 *If $\{X(t)\}$ satistifies either Assumption 3.1, 4.1, or 3.2, then the LSE $\widehat{\theta}$ and the ALSE $\widetilde{\theta}$ are both strongly consistent estimators of θ^0, that is,*

$$\widehat{\theta} \xrightarrow{a.s.} \theta^0 \quad and \quad \widetilde{\theta} \xrightarrow{a.s.} \theta^0. \tag{4.2}$$

The following lemmas are required to prove Theorem 4.1.

Lemma 4.1 *Let $S_{C_1,K} = \{\theta; \theta = (A, B, \theta), |\theta - \theta^0| \ge 3C_1, |A| \le K, |B| \le K\}$. If for any $C_1 > 0$ and for some $K < \infty$,*

$$\liminf_{n \to \infty} \inf_{\theta \in S_{C_1,K}} \frac{1}{n} \left[Q(\theta) - Q(\theta^0) \right] > 0 \quad a.s., \tag{4.3}$$

then $\widehat{\theta}$ is a strongly consistent estimator of θ^0.

Lemma 4.2 *If $\{X(t)\}$ satisfies either Assumption 3.1, 4.1, or 3.2, then*

$$\sup_{\omega \in (0,\pi)} \left| \frac{1}{n} \sum_{t=1}^{n} X(t) \cos(\omega t) \right| \to 0 \quad a.s. \quad as \ n \to \infty. \tag{4.4}$$

Corollary 4.1

$$\sup_{\omega \in (0,\pi)} \left| \frac{1}{n^{k+1}} \sum_{t=1}^{n} t^k X(t) \cos(\omega t) \right| \to 0 \quad a.s., \quad for \ k = 1, 2 \ldots$$

The result is true for sine functions also.

Lemma 4.3 *Write* $S_{C_2} = \{\omega : |\omega - \omega^0| > C_2\}$, *for any* $C_2 > 0$. *If for some* $C_2 > 0$,

$$\overline{\lim_{S_{C_2}}} \sup \frac{1}{n} \left[I(\omega) - I(\omega^0) \right] < 0 \quad a.s.,$$

then $\widetilde{\omega}$, *the ALSE of* ω^0, *converges to* ω^0 *a.s. as* $n \to \infty$.

Lemma 4.4 *Suppose* $\widetilde{\omega}$ *is the ALSE of* ω^0. *Then* $n(\widetilde{\omega} - \omega^0) \to 0$ *a.s. as* $n \to \infty$.

Lemma 4.1 provides a sufficient condition for $\widehat{\boldsymbol{\theta}}$ to be strongly consistent, whereas Lemma 4.3 gives a similar condition for $\widetilde{\omega}$, the ALSE of ω^0. Lemma 4.2 is used to verify conditions given in Lemmas 4.1 and 4.3. Lemma 4.4 is required to prove the strong consistency of the ALSEs of the amplitudes, $\widetilde{A}$ and $\widetilde{B}$.

The consistency results of the LSEs and the ALSEs are stated in Theorem 4.1 in concise form and it is proved in two steps. First, the proof of the strong consistency of $\widehat{\boldsymbol{\theta}}$, the LSE of $\boldsymbol{\theta}$ and next the proof of the strong consistency of $\widetilde{\boldsymbol{\theta}}$, the ALSE of $\boldsymbol{\theta}$. The proofs of Lemmas 4.1–4.4, required to prove Theorem 4.1, are given in Appendix A. We prove Lemma 4.1. The proof of Lemma 4.3 is similar to Lemma 4.1 and so it is omitted. Lemma 4.2 is proved separately under Assumptions 4.1 and 3.2.

4.3.1 Proof of the Strong Consistency of $\widehat{\boldsymbol{\theta}}$, the LSE of $\boldsymbol{\theta}$

In this proof, we denote $\widehat{\boldsymbol{\theta}}$ by $\widehat{\boldsymbol{\theta}}_n$ to write explicitly that $\widehat{\boldsymbol{\theta}}$ depends on n. If $\widehat{\boldsymbol{\theta}}_n$ is not consistent for $\boldsymbol{\theta}^0$, then either case 4.1 or case 4.2 occurs.

Case 4.1 For all subsequences $\{n_k\}$ of $\{n\}$, $|\widehat{A}_n| + |\widehat{B}_n| \to \infty$. This implies $\left[Q(\widehat{\boldsymbol{\theta}}_{n_k}) - Q(\boldsymbol{\theta}^0) \right] / n_k \to \infty$. At the same time, $\widehat{\boldsymbol{\theta}}_{n_k}$ is the LSE of $\boldsymbol{\theta}^0$ at $n = n_k$, therefore $Q(\widehat{\boldsymbol{\theta}}_{n_k}) - Q(\boldsymbol{\theta}^0) < 0$. This leads to a contradiction.

Case 4.2 For at least one subsequence $\{n_k\}$ of $\{n\}$, $\widehat{\boldsymbol{\theta}}_{n_k} \in S_{C_1, K}$ for some $C_1 > 0$ and for a $0 < K < \infty$. Write $\left[Q(\boldsymbol{\theta}) - Q(\boldsymbol{\theta}^0) \right] / n = f_1(\boldsymbol{\theta}) + f_2(\boldsymbol{\theta})$, where

$$f_1(\boldsymbol{\theta}) = \frac{1}{n} \sum_{t=1}^{n} \left[A^0 \cos(\omega^0 t) - A \cos(\omega t) + B^0 \sin(\omega^0 t) - B \sin(\omega t) \right]^2,$$

$$f_2(\boldsymbol{\theta}) = \frac{1}{n} \sum_{t=1}^{n} X(t) \left[A^0 \cos(\omega^0 t) - A \cos(\omega t) + B^0 \sin(\omega^0 t) - B \sin(\omega t) \right].$$

Define sets $S_{C_1,K}^j = \{\boldsymbol{\theta} : |\theta_j - \theta_j^0| > C_1, |A| \le K, |B| \le K\}$ for $j = 1, 2, 3$, where θ_j is the jth element of $\boldsymbol{\theta}$, that is, $\theta_1 = A$, $\theta_2 = B$, and $\theta_3 = \omega$ and θ_j^0 is the true values of θ_j. Then $S_{C_1,K} \subset S_{C_1,K}^1 \cup S_{C_1,K}^2 \cup S_{C_1,K}^3 = S$, say and

$$\liminf_{n \to \infty} \inf_{S_{C_1,K}} \frac{1}{n} \left[Q(\boldsymbol{\theta}) - Q(\boldsymbol{\theta}^0) \right] \ge \liminf_{n \to \infty} \inf_{S} \frac{1}{n} \left[Q(\boldsymbol{\theta}) - Q(\boldsymbol{\theta}^0) \right].$$

Using Lemma 4.2, $\lim_{n \to \infty} f_2(\boldsymbol{\theta}) = 0$ a.s. Then

$$\liminf_{n \to \infty} \inf_{S_{C_1,K}^j} \frac{1}{n} \left[Q(\boldsymbol{\theta}) - Q(\boldsymbol{\theta}^0) \right] = \liminf_{n \to \infty} \inf_{S_{C_1,K}^j} f_1(\boldsymbol{\theta}) > 0 \text{ a.s. for } j = 1, \ldots, 4,$$

$$\Rightarrow \liminf_{n \to \infty} \inf_{S_{C_1,K}} \frac{1}{n} \left[Q(\boldsymbol{\theta}) - Q(\boldsymbol{\theta}^0) \right] > 0 \ a.s.$$

Therefore, for $j = 1$,

$$\liminf_{n \to \infty} \inf_{S_{C_1,K}^j} f_1(\boldsymbol{\theta})$$

$$= \liminf_{n \to \infty} \inf_{|A-A^0|>C_1} \frac{1}{n} \sum_{t=1}^{n} \left[\left\{ A^0 \cos(\omega^0 t) - A \cos(\omega t) \right\}^2 + \left\{ B^0 \sin(\omega^0 t) - B \sin(\omega t) \right\}^2 \right.$$

$$\left. + 2 \left\{ A^0 \cos(\omega^0 t) - A \cos(\omega t) \right\} \left\{ B^0 \sin(\omega^0 t) - B \sin(\omega t) \right\} \right]$$

$$= \liminf_{n \to \infty} \inf_{|A-A^0|>C_1} \frac{1}{n} \sum_{t=1}^{n} \left[\left\{ A^0 \cos(\omega^0 t) - A \cos(\omega t) \right\}^2 > \frac{1}{2} C_1^2 > 0 \text{ a.s.} \right.$$

We have used trigonometric result (2.5) here. Similarly, the inequality holds for $j = 2, 3$. Therefore, using Lemma 4.1, we say that $\widehat{\boldsymbol{\theta}}$ is a strongly consistent estimator of $\boldsymbol{\theta}$. □

4.3.2 Proof of Strong Consistency of $\widetilde{\boldsymbol{\theta}}$, the ALSE of $\boldsymbol{\theta}$

We first prove the consistency of $\widetilde{\omega}$, the ALSE of ω, and then provide the proof of the linear parameter estimators. Consider

$$\frac{1}{n}\left[I(\omega) - I(\omega^0)\right]$$

$$= \frac{1}{n^2}\left[\left|\sum_{t=1}^{n} y(t)e^{-i\omega t}\right|^2 - \left|\sum_{t=1}^{n} y(t)e^{-i\omega^0 t}\right|^2\right]$$

$$= \frac{1}{n^2}\left[\left\{\sum_{t=1}^{n} y(t)\cos(\omega t)\right\}^2 + \left\{\sum_{t=1}^{n} y(t)\sin(\omega t)\right\}^2\right.$$

$$\left. - \left\{\sum_{t=1}^{n} y(t)\cos(\omega^0 t)\right\}^2 - \left\{\sum_{t=1}^{n} y(t)\sin(\omega^0 t)\right\}^2\right]$$

$$= \left\{\sum_{t=1}^{n}\left(A^0\cos(\omega^0 t) + B^0\sin(\omega^0 t) + X(t)\right)\cos(\omega t)\right\}^2$$

$$+ \left\{\sum_{t=1}^{n}\left(A^0\cos(\omega^0 t) + B^0\sin(\omega^0 t) + X(t)\right)\sin(\omega t)\right\}^2$$

$$- \left\{\sum_{t=1}^{n}\left(A^0\cos(\omega^0 t) + B^0\sin(\omega^0 t) + X(t)\right)\cos(\omega^0 t)\right\}^2$$

$$- \left\{\sum_{t=1}^{n}\left(A^0\cos(\omega^0 t) + B^0\sin(\omega^0 t) + X(t)\right)\sin(\omega^0 t)\right\}^2.$$

Using Lemma 4.2, the terms of the form $\overline{\lim}\sup_{\omega\in S_{C_2}}(1/n)\sum_{t=1}^{n} X(t)\cos(\omega t) = 0$ a.s. and using trigonometric identities (2.5–2.7), we have

$$\overline{\lim}\sup_{\omega\in S_{C_2}}\frac{1}{n}\left[I(\omega) - I(\omega^0)\right]$$

$$= -\lim_{n\to\infty}\left\{\frac{1}{n}\sum_{t=1}^{n} A^0\cos^2(\omega^0 t)\right\}^2 - \lim_{n\to\infty}\left\{\frac{1}{n}\sum_{t=1}^{n} B^0\sin^2(\omega^0 t)\right\}^2$$

$$= -\frac{1}{4}(A^{0^2} + B^{0^2}) < 0 \quad a.s.$$

Therefore, using Lemma 4.3, $\widetilde{\omega} \to \omega^0$ a.s.

We need Lemma 4.4 to prove that $\widetilde{A}$ and $\widetilde{B}$ are strongly consistent. Observe that

$$\widetilde{A} = \frac{2}{n}\sum_{t=1}^{n} y(t)\cos(\widetilde{\omega}t) = \frac{2}{n}\sum_{t=1}^{n}\left(A^0\cos(\omega^0 t) + B^0\sin(\omega^0 t) + X(t)\right)\cos(\widetilde{\omega}t).$$

Using Lemma 4.2, $(2/n)\sum_{t=1}^{n} X(t)\cos(\widetilde{\omega}t) \to 0$. Expand $\cos(\widetilde{\omega}t)$ by Taylor series around ω^0.

$$\widetilde{A} = \frac{2}{n} \sum_{t=1}^{n} \left(A^0 \cos(\omega^0 t) + B^0 \sin(\omega^0 t) \right) \left[\cos(\omega^0 t) - t(\widetilde{\omega} - \omega^0) \sin(\omega t) \right] \to A^0 \quad a.s.$$

using Lemma 4.4 and trigonometric results (2.5–2.7).   □

Remark 4.1 The proof of the consistency of LSE and ALSE of $\widehat{\theta}$ for model (4.1) is extensively discussed. The consistency of the LSE of the parameter vector of the multiple sinusoidal model with $p > 1$ follows similarly as the consistency of $\widehat{\theta}$ and $\widetilde{\theta}$ of this section.

4.4 Asymptotic Distribution of LSE and ALSE of θ

This section discusses the asymptotic distribution of LSE and ALSE of $\widehat{\theta}$ under different error assumptions. We discuss the asymptotic distribution of $\widehat{\theta}$, the LSE of θ under Assumption 3.2. Then we consider the case under Assumption 4.2. Under Assumption 3.2, the asymptotic distribution of $\widehat{\theta}$, as well as $\widetilde{\theta}$, is multivariate normal, whereas under Assumption 4.2, it is distributed as multivariate symmetric stable. First we discuss the asymptotic distribution of $\widehat{\theta}$ in both the cases. Then the asymptotic distribution of $\widetilde{\theta}$ is shown to be same as that of $\widehat{\theta}$ by proving the asymptotic equivalence of $\widehat{\theta}$ and $\widetilde{\theta}$.

4.4.1 Asymptotic Distribution of $\widehat{\theta}$ Under Assumption 3.2

Assume that $\{X(t)\}$ satisfies Assumption 3.2. Let $Q'(\theta)$ be a 1×3 vector of first derivative and $Q''(\theta)$, a 3×3 matrix of second derivatives of $Q(\theta)$, that is,

$$Q'(\theta) = \left(\frac{\partial Q(\theta)}{\partial A}, \frac{\partial Q(\theta)}{\partial B}, \frac{\partial Q(\theta)}{\partial \omega} \right), \tag{4.5}$$

$$Q''(\theta) = \begin{pmatrix} \frac{\partial^2 Q(\theta)}{\partial A^2} & \frac{\partial^2 Q(\theta)}{\partial A \partial B} & \frac{\partial^2 Q(\theta)}{\partial A \partial \omega} \\ \frac{\partial^2 Q(\theta)}{\partial B \partial A} & \frac{\partial^2 Q(\theta)}{\partial B^2} & \frac{\partial^2 Q(\theta)}{\partial B \partial \omega} \\ \frac{\partial^2 Q(\theta)}{\partial \omega \partial A} & \frac{\partial^2 Q(\theta)}{\partial \omega \partial B} & \frac{\partial^2 Q(\theta)}{\partial \omega^2} \end{pmatrix}. \tag{4.6}$$

The elements of $Q'(\theta)$ and $Q''(\theta)$ are

$$\frac{\partial Q(\boldsymbol{\theta})}{\partial A} = -2 \sum_{t=1}^{n} X(t)\cos(\omega t), \qquad \frac{\partial Q(\boldsymbol{\theta})}{\partial B} = -2 \sum_{t=1}^{n} X(t)\sin(\omega t),$$

$$\frac{\partial Q(\boldsymbol{\theta})}{\partial \omega} = 2 \sum_{t=1}^{n} t X(t) \left[A\sin(\omega t) - B\cos(\omega t) \right],$$

$$\frac{\partial^2 Q(\boldsymbol{\theta})}{\partial A^2} = 2 \sum_{t=1}^{n} \cos^2(\omega t), \qquad \frac{\partial^2 Q(\boldsymbol{\theta})}{\partial B^2} = 2 \sum_{t=1}^{n} \sin^2(\omega t),$$

$$\frac{\partial^2 Q(\boldsymbol{\theta})}{\partial A \partial B} = 2 \sum_{t=1}^{n} \cos(\omega t)\sin(\omega t),$$

$$\frac{\partial^2 Q(\boldsymbol{\theta})}{\partial A \partial \omega} = -2 \sum_{t=1}^{n} t\cos(\omega t) \left[A\sin(\omega t) - B\cos(\omega t) \right] + 2 \sum_{t=1}^{n} t X(t)\sin(\omega t),$$

$$\frac{\partial^2 Q(\boldsymbol{\theta})}{\partial B \partial \omega} = -2 \sum_{t=1}^{n} t\sin(\omega t) \left[A\sin(\omega t) - B\cos(\omega t) \right] - 2 \sum_{t=1}^{n} t X(t)\cos(\omega t),$$

$$\frac{\partial^2 Q(\boldsymbol{\theta})}{\partial \omega^2} = 2 \sum_{t=1}^{n} t^2 \left[A\sin(\omega t) - B\cos(\omega t) \right]^2$$

$$+ 2 \sum_{t=1}^{n} t^2 X(t) \left[A\cos(\omega t) + B\sin(\omega t) \right].$$

Consider a 3×3 diagonal matrix $\mathbf{D} = \mathrm{diag}\left\{ n^{-1/2}, n^{-1/2}, n^{-3/2} \right\}$. Expanding $Q'(\widehat{\boldsymbol{\theta}})$ around $\boldsymbol{\theta}^0$ using Taylor series expansion

$$Q'(\widehat{\boldsymbol{\theta}}) - Q'(\boldsymbol{\theta}^0) = (\widehat{\boldsymbol{\theta}} - \boldsymbol{\theta}^0) Q''(\bar{\boldsymbol{\theta}}), \qquad (4.7)$$

where $\bar{\boldsymbol{\theta}}$ is a point on the line joining $\widehat{\boldsymbol{\theta}}$ and $\boldsymbol{\theta}^0$. As $\widehat{\boldsymbol{\theta}}$ is the LSE of $\boldsymbol{\theta}^0$, $Q'(\widehat{\boldsymbol{\theta}}) = 0$. Also $\widehat{\boldsymbol{\theta}} \overset{a.s.}{\to} \boldsymbol{\theta}^0$ using Theorem 4.1. Because $Q(\boldsymbol{\theta})$ is a continuous function of $\boldsymbol{\theta}$, we have

$$\lim_{n \to \infty} \mathbf{D} Q''(\bar{\boldsymbol{\theta}})\mathbf{D} = \lim_{n \to \infty} \mathbf{D} Q''(\boldsymbol{\theta}^0)\mathbf{D} = \begin{pmatrix} 1 & 0 & \frac{1}{2}B^0 \\ 0 & 1 & -\frac{1}{2}A^0 \\ \frac{1}{2}B^0 & -\frac{1}{2}A^0 & \frac{1}{3}(A^{0^2} + B^{0^2}) \end{pmatrix} = \boldsymbol{\Sigma}, \quad \text{say.}$$

$$(4.8)$$

Therefore, (4.7) can be written as

$$(\widehat{\boldsymbol{\theta}} - \boldsymbol{\theta}^0)\mathbf{D}^{-1} = - \left[Q'(\boldsymbol{\theta}^0)\mathbf{D} \right] \left[\mathbf{D} Q''(\bar{\boldsymbol{\theta}})\mathbf{D} \right]^{-1}, \qquad (4.9)$$

since $\mathbf{D} Q''(\bar{\boldsymbol{\theta}})\mathbf{D}$ is an invertible matrix a.e. for large n. Using a central limit theorem of stochastic processes, Fuller [17], it follows that $Q'(\boldsymbol{\theta}^0)\mathbf{D}$ tends to a 3-variate normal distribution with mean vector zero and variance covariance matrix equal to

$2\sigma^2 c(\omega^0) \boldsymbol{\Sigma}$, where

$$c(\omega^0) = \left| \sum_{j=0}^{\infty} a(j) e^{-ij\omega^0} \right|^2 = \left[\sum_{j=0}^{\infty} a(j) \cos(j\omega^0) \right]^2 + \left[\sum_{j=0}^{\infty} a(j) \sin(j\omega^0) \right]^2.$$

(4.10)

Therefore, $(\widehat{\boldsymbol{\theta}} - \boldsymbol{\theta}^0) \mathbf{D}^{-1} \xrightarrow{d} \mathcal{N}_3(\mathbf{0}, 2\sigma^2 c(\omega^0) \boldsymbol{\Sigma}^{-1})$ and we can state the asymptotic distribution in the following theorem.

Theorem 4.2 *Under Assumption 3.2, the limiting distribution of* $(\{n^{\frac{1}{2}}(\widehat{A} - A^0),$ $n^{\frac{1}{2}}(\widehat{B} - B^0), n^{\frac{3}{2}}(\widehat{\omega} - \omega^0)\})$ *as* $n \to \infty$ *is a 3-variate normal distribution with mean vector zero and dispersion matrix* $2\sigma^2 c(\omega^0)\boldsymbol{\Sigma}^{-1}$, *where* $c(\omega^0)$ *is defined in (4.10) and* $\boldsymbol{\Sigma}^{-1}$ *has the following form:*

$$\boldsymbol{\Sigma}^{-1} = \frac{1}{A^{0^2} + B^{0^2}} \begin{bmatrix} A^{0^2} + 4B^{0^2} & -3A^0 B^0 & -3B^0 \\ -3A^0 B^0 & 4A^{0^2} + B^{0^2} & 3A^0 \\ -3B^0 & 3A^0 & 6 \end{bmatrix}. \quad (4.11)$$

Remark 4.2

1. The diagonal entries of matrix $\mathbf{D}$ correspond to the rates of convergence of $\widehat{A}, \widehat{B}$, and $\widehat{\omega}$, respectively. Therefore, $\widehat{A} - A^0 = O_p(n^{-1/2})$, $\widehat{B} - B^0 = O_p(n^{-1/2})$, and $\widehat{\omega} - \omega^0 = O_p(n^{-3/2})$.
2. Instead of Assumption 3.2, if $\{X(t)\}$ only satisfies Assumption 3.1, then $a(j) = 0$, for all $j \neq 0$ and $a(0) = 1$ in the derivation discussed above and so $c(\omega^0) = 1$. Therefore, in such a situation $(\widehat{\boldsymbol{\theta}} - \boldsymbol{\theta}^0)\mathbf{D}^{-1} \xrightarrow{d} \mathcal{N}_3(\mathbf{0}, 2\sigma^2 \boldsymbol{\Sigma}^{-1})$.
3. Observe that $(\sigma^2/2\pi)c(\omega) = f(\omega)$, where $f(\omega)$ is the spectral density function of the error process $\{X(t)\}$ under Assumption 3.2.

4.4.2 Asymptotic Distribution of $\widehat{\theta}$ Under Assumption 4.2

In this section, the asymptotic distribution of $\widehat{\boldsymbol{\theta}}$ under Assumption 4.2 is developed, that is, $\{X(t)\}$ is a sequence of i.i.d. symmetric stable random variables with stability index α and scale parameter σ (see Nandi et al. [14]).

Define two diagonal matrices of order 3×3 as follows:

$$\mathbf{D}_1 = \text{diag}\left\{ n^{-\frac{1}{\alpha}}, n^{-\frac{1}{\alpha}}, n^{-\frac{1+\alpha}{\alpha}} \right\}, \qquad \mathbf{D}_2 = \text{diag}\left\{ n^{-\frac{\alpha-1}{\alpha}}, n^{-\frac{\alpha-1}{\alpha}}, n^{-\frac{2\alpha-1}{\alpha}} \right\}. \quad (4.12)$$

Note that $\mathbf{D}_1 \mathbf{D}_2 = \mathbf{D}^2$. Also if $\alpha = 2$, it corresponds to normal distribution and in that case $\mathbf{D}_1 = \mathbf{D}_2 = \mathbf{D}$. Using the same argument as in (4.8), we have

$$\lim_{n\to\infty} \mathbf{D}_2 Q''(\bar{\theta})\mathbf{D}_1 = \lim_{n\to\infty} \mathbf{D}_2 Q''(\theta^0)\mathbf{D}_1 = \Sigma. \tag{4.13}$$

Similarly as in (4.9), we write

$$(\widehat{\theta} - \theta^0)\mathbf{D}_2^{-1} = -\left[Q'(\theta^0)\mathbf{D}_1\right]\left[\mathbf{D}_2 Q''(\bar{\theta})\mathbf{D}_1\right]^{-1}. \tag{4.14}$$

To find the distribution of $Q'(\theta^0)\mathbf{D}_1$, write

$$Q'(\theta^0)\mathbf{D}_1 = \left[-\frac{2}{n^{\frac{1}{\alpha}}} \sum_{t=1}^{n} X(t)\cos(\omega^0 t), \ -\frac{2}{n^{\frac{1}{\alpha}}} \sum_{t=1}^{n} X(t)\sin(\omega^0 t), \right.$$

$$\left. \frac{2}{n^{\frac{1+\alpha}{\alpha}}} \sum_{t=1}^{n} t X(t) \left[A^0 \sin(\omega^0 t) - B^0 \cos(\omega^0 t) \right] \right]$$

$$= \left(Z_n^1, Z_n^2, Z_n^3 \right), \quad \text{(say)}. \tag{4.15}$$

Then the joint characteristic function of (Z_n^1, Z_n^2, Z_n^3) is

$$\phi_n(\mathbf{t}) = E\exp\{i(t_1 Z_n^1 + t_2 Z_n^2 + t_3 Z_n^3)\} = E\exp\left\{ i\frac{2}{n^{1/\alpha}} \sum_{j=1}^{n} X(j)K_{\mathbf{t}}(j) \right\}, \tag{4.16}$$

where

$$K_{\mathbf{t}}(j) = -t_1\cos(\omega^0 j) - t_2\sin(\omega^0 j) + \frac{jt_3}{n}\left\{ A^0\sin(\omega^0 j) - B^0\cos(\omega^0 j) \right\}. \tag{4.17}$$

Since $\{X(t)\}$ is a sequence of i.i.d. random variables

$$\phi_n(t) = \prod_{j=1}^{n} \exp\{-2^\alpha\sigma^\alpha \frac{1}{n}|K_{\mathbf{t}}(j)|^\alpha\} = \exp\{-2^\alpha\sigma^\alpha\frac{1}{n}\sum_{j=1}^{n}|K_{\mathbf{t}}(j)|^\alpha\}. \tag{4.18}$$

Nandi et al. [14] argued that $(1/n)\sum_{j=1}^{n}|K_{\mathbf{t}}(j)|^\alpha$ converges, based on extensive numerical experiments. Assuming that it converges, it is proved in Nandi et al. [14] that it converges to a non-zero limit for $\mathbf{t} \neq \mathbf{0}$. The proof is given in Appendix B. Suppose

$$\lim_{n\to\infty} \frac{1}{n}\sum_{j=1}^{n}|K_{\mathbf{t}}(j)|^\alpha = \tau_{\mathbf{t}}(A^0, B^0, \omega^0, \alpha). \tag{4.19}$$

Therefore, the limiting characteristic function is

$$\lim_{n \to \infty} \phi_n(\mathbf{t}) = e^{-2^\alpha \sigma^\alpha \tau_t(A^0, B^0, \omega^0, \alpha)}, \qquad (4.20)$$

which indicates that even if $n \to \infty$, any linear combination of Z_n^1, Z_n^2, and Z_n^3, follows a $S\alpha S$ distribution. Using Theorem 2.1.5 of Samorodnitsky and Taqqu [18], it follows that:

$$\lim_{n \to \infty} \left[Q'(\theta^0)\mathbf{D_1} \right] \left[\mathbf{D_2} Q''(\bar{\theta})\mathbf{D_1} \right]^{-1} \qquad (4.21)$$

converges to a symmetric α stable random vector in $\mathbb{R}^3$ with characteristic function $\phi(\mathbf{t}) = \exp\{-2^\alpha \sigma^\alpha \tau_\mathbf{u}(A^0, B^0, \omega^0, \alpha)\}$, where $\tau_\mathbf{u}$ is defined through (4.19) replacing $\mathbf{t}$ by $\mathbf{u}$. The vector $\mathbf{u}$ is defined as a function of $\mathbf{t}$ as $\mathbf{u} = (u_1, u_2, u_3)$ with

$$u_1(t_1, t_2, t_3, A^0, B^0) = \left[(A^{0^2} + 4B^{0^2})t_1 - 3A^0 B^0 t_2 - 6B^0 t_3 \right] \frac{1}{A^{0^2} + B^{0^2}},$$

$$u_2(t_1, t_2, t_3, A^0, B^0) = \left[-3A^0 B^0 t_1 + (4A^{0^2} + B^{0^2})t_2 + 6A^0 t_3 \right] \frac{1}{A^{0^2} + B^{0^2}},$$

$$u_3(t_1, t_2, t_3, A^0, B^0) = \left[-6B^0 t_1 + 6A^0 t_2 + 12t_3 \right] \frac{1}{A^{0^2} + B^{0^2}}.$$

Therefore, we have the following theorem.

Theorem 4.3 *Under Assumption 4.2,* $(\widehat{\theta} - \theta^0)\mathbf{D_2}^{-1} = \left(n^{\frac{\alpha-1}{\alpha}}(\widehat{A} - A^0), n^{\frac{\alpha-1}{\alpha}}(\widehat{B} - B^0), n^{\frac{2\alpha-1}{\alpha}}(\omega^0 - \omega) \right)$ *converges to a multivariate symmetric stable distribution in* $\mathbb{R}^3$ *having characteristic function equal to* $\phi(\mathbf{t})$.

4.4.3 Asymptotic Equivalence of LSE $\widehat{\theta}$ and ALSE θ

In this section, it is shown that the asymptotic distribution of $\widetilde{\theta}$, the ALSE of θ, is equivalent to that of $\widehat{\theta}$ for large n. We have presented the asymptotic distribution of $\widehat{\theta}$ in two cases: (i) $\{X(t)\}$ is a sequence of i.i.d. symmetric stable with stability index $1 < \alpha < 2$ and scale parameter σ, and (ii) $\{X(t)\}$ is a stationary linear process, such that it can be expressed as (3.2) with absolute summable coefficients. In both the cases, the asymptotic distribution of $\widetilde{\theta}$ is same as the LSE, $\widehat{\theta}$ and stated in the following theorem.

Theorem 4.4 *Under Assumption 4.2, the asymptotic distribution of* $(\widetilde{\theta} - \theta^0)\mathbf{D_2}^{-1}$ *is same as that of* $(\widehat{\theta} - \theta^0)\mathbf{D_2}^{-1}$. *Similarly, under Assumption 3.2 the asymptotic distribution of* $(\widetilde{\theta} - \theta^0)\mathbf{D}^{-1}$ *is same as that of* $(\widehat{\theta} - \theta^0)\mathbf{D}^{-1}$.

Proof Observe that under Assumption 4.2

$$\frac{1}{n} Q(\theta)$$

$$= \frac{1}{n} \sum_{t=1}^{n} y(t)^2 - \frac{2}{n} \sum_{t=1}^{n} y(t) \{A \cos(\omega t) + B \sin(\omega t)\}$$

$$+ \frac{1}{n} \sum_{t=1}^{n} (A \cos(\omega t) + B \sin(\omega t))^2$$

$$= \frac{1}{n} \sum_{t=1}^{n} y(t)^2 - \frac{2}{n} \sum_{t=1}^{n} y(t) \{A \cos(\omega t) + B \sin(\omega t)\} + \frac{1}{2} \left(A^2 + B^2\right) + O\left(\frac{1}{n}\right)$$

$$= C - \frac{1}{n} J(\theta) + O\left(\frac{1}{n}\right), \tag{4.22}$$

where

$$C = \frac{1}{n} \sum_{t=1}^{n} y(t)^2 \quad \text{and}$$

$$\frac{1}{n} J(\theta) = \frac{2}{n} \sum_{t=1}^{n} y(t) \{A \cos(\omega t) + B \sin(\omega t)\} - \frac{1}{2} \left(A^2 + B^2\right).$$

Write $J'(\theta)/n = \left(\frac{\partial J(\theta)}{\partial A}/n, \frac{\partial J(\theta)}{\partial B}/n, \frac{\partial J(\theta)}{\partial \omega}/n\right)$, then at θ^0

$$\frac{1}{n} \frac{\partial J(\theta^0)}{\partial A} = \frac{2}{n} \sum_{t=1}^{n} y(t) \cos(\omega^0 t) - A^0$$

$$= \frac{2}{n} \sum_{t=1}^{n} \left\{A^0 \cos(\omega^0 t) + B^0 \sin(\omega^0 t) + X(t)\right\} \cos(\omega^0 t) - A^0$$

$$= \frac{2}{n} \sum_{t=1}^{n} X(t) \cos(\omega^0 t) + \frac{2A^0}{n} \sum_{t=1}^{n} \cos^2(\omega^0 t)$$

$$+ \frac{2B^0}{n} \sum_{t=1}^{n} \sin(\omega^0 t) \cos(\omega^0 t) - A^0$$

$$= \frac{2}{n} \sum_{t=1}^{n} X(t) \cos(\omega^0 t) + A^0 + O\left(\frac{1}{n}\right) - A^0,$$

$$= \frac{2}{n} \sum_{t=1}^{n} X(t) \cos(\omega^0 t) + O\left(\frac{1}{n}\right).$$

Similarly $(1/n) \dfrac{\partial J(\theta^0)}{\partial B} = (2/n) \displaystyle\sum_{t=1}^{n} X(t) \sin(\omega^0 t) + O\left(1/n\right)$, and

$$\frac{1}{n}\frac{\partial J(\theta^0)}{\partial \omega} = \frac{2}{n}\sum_{t=1}^{n} tX(t)\left\{-A^0\sin(\omega^0 t) + B^0\cos(\omega^0 t)\right\} + O(1). \qquad (4.23)$$

Comparing $Q'(\theta^0)/n$ and $J'(\theta^0)/n$, we have

$$\frac{1}{n}Q'(\theta^0) = -\frac{1}{n}J'(\theta^0) + \begin{bmatrix} O\left(1/n\right) \\ O\left(1/n\right) \\ O(1) \end{bmatrix}^T \Rightarrow Q'(\theta^0) = -J'(\theta^0) + \begin{bmatrix} O(1) \\ O(1) \\ O(n) \end{bmatrix}^T.$$
$$(4.24)$$

Note that $\widetilde{A} = \widetilde{A}(\omega)$ and $\widetilde{B} = \widetilde{B}(\omega)$, therefore, at $(\widetilde{A}, \widetilde{B}, \omega)$

$$J(\widetilde{A}, \widetilde{B}, \omega)$$
$$= 2\sum_{t=1}^{n} y(t)\left[\left\{\frac{2}{n}\sum_{k=1}^{n} y(k)\cos(\omega k)\right\}\cos(\omega t) + \left\{\frac{2}{n}\sum_{k=1}^{n} y(k)\sin(\omega k)\right\}\sin(\omega t)\right]$$
$$- \frac{n}{2}\left[\left\{\frac{2}{n}\sum_{t=1}^{n} y(t)\cos(\omega t)\right\}^2 + \left\{\frac{2}{n}\sum_{t=1}^{n} y(t)\sin(\omega t)\right\}^2\right]$$
$$= \frac{2}{n}\left\{\sum_{t=1}^{n} y(t)\cos(\omega t)\right\}^2 + \frac{2}{n}\left\{\sum_{t=1}^{n} y(t)\sin(\omega t)\right\}^2 = \frac{2}{n}\left|\sum_{t=1}^{n} y(t)e^{-i\omega t}\right|^2 = I(\omega).$$

Hence, the estimator of θ^0, which maximizes $J(\theta)$, is equivalent to $\widetilde{\theta}$, the ALSE of θ^0. Thus, for the ALSE $\widetilde{\theta}$, in terms of $J(\theta)$, is

$$(\widetilde{\theta} - \theta^0) = -J'(\theta^0)\left[J''(\bar{\theta})\right]^{-1}$$
$$\Rightarrow (\widetilde{\theta} - \theta^0)\mathbf{D_2}^{-1} = -\left[J'(\theta^0)\mathbf{D_1}\right]\left[\mathbf{D_2}J''(\bar{\theta})\mathbf{D_1}\right]^{-1}$$
$$= -\left[\left(-Q'(\theta^0) + \begin{bmatrix} O(1) \\ O(1) \\ O(n) \end{bmatrix}^T\right)\mathbf{D_1}\right]\left[\mathbf{D_2}J''(\bar{\theta})\mathbf{D_1}\right]^{-1}. \qquad (4.25)$$

The matrices $\mathbf{D_1}$ and $\mathbf{D_2}$ are same as defined in (4.12). One can show similarly as in (4.8) and (4.13), that

$$\lim_{n\to\infty}\left[\mathbf{D_2}J''(\bar{\theta})\mathbf{D_1}\right] = \lim_{n\to\infty}\left[\mathbf{D_2}J''(\theta^0)\mathbf{D_1}\right] = -\Sigma = -\lim_{n\to\infty}\left[\mathbf{D_2}Q''(\theta^0)\mathbf{D_1}\right].$$
$$(4.26)$$

Using (4.14) and (4.26) in (4.25), we have

$$(\widetilde{\theta} - \theta^0)\mathbf{D}_2^{-1} = -\left[Q'(\theta^0)\mathbf{D}_1\right]\left[\mathbf{D}_2 Q''(\bar{\theta})\mathbf{D}_1\right]^{-1} + \begin{pmatrix} O(1) \\ O(1) \\ O(n) \end{pmatrix}^T \mathbf{D}_1\left[\mathbf{D}_2 J''(\bar{\theta})\mathbf{D}_1\right]^{-1}$$

$$= (\widehat{\theta} - \theta^0)\mathbf{D}_2^{-1} + \begin{pmatrix} O(1) \\ O(1) \\ O(n) \end{pmatrix}^T \mathbf{D}_1\left[\mathbf{D}_2 J''(\bar{\theta})\mathbf{D}_1\right]^{-1}.$$

Since $\mathbf{D}_2 J''(\theta^0)\mathbf{D}_1$ is an invertible matrix a.e. for large n and $\lim_{n\to\infty} \begin{pmatrix} O(1) \\ O(1) \\ O(n) \end{pmatrix}^T$
$\mathbf{D}_1 = \mathbf{0}$, it follows that LSE, $\widehat{\theta}$ and ALSE, $\widetilde{\theta}$ of θ^0 of model (4.1) are asymptotically equivalent in distribution. Therefore, asymptotic distribution of $\widetilde{\theta}$ is same as that of $\widehat{\theta}$.

Under Assumption 3.2, instead of Assumption 4.2, (4.25) follows similarly by replacing $\mathbf{D}_1 = \mathbf{D}_2 = \mathbf{D}$. This is the case corresponding to $\alpha = 2$, so that the second moment is finite. Similarly (4.26), and equivalence follow. □

4.5 Superefficient Frequency Estimator

In this section, we discuss the theoretical results behind the superefficient algorithm proposed by Kundu et al. [16]. This method modifies the widely used Newton–Raphson iterative method. In the previous section, we have seen that the least squares method estimates the frequency with convergence rate $O_p(n^{-3/2})$ and once the frequency is estimated with $O_p(n^{-3/2})$, the linear parameters can be estimated efficiently with the rate of convergence $O_p(n^{-1/2})$. The modified Newton-Raphson method estimates the frequency with the same rate of convergence as the LSE and the asymptotic variance is smaller than that of the LSE.

The superefficient frequency estimator of ω maximizes $S(\omega)$, where $S(\omega)$ is defined in (3.55) in Sect. 3.17. Suppose $\widehat{\omega}$ maximizes $S(\omega)$, then the estimators of A and B are obtained using the separable regression technique as

$$(\widehat{A} \ \widehat{B})^T = (\mathbf{Z}(\widehat{\omega})^T \mathbf{Z}(\widehat{\omega}))^{-1}\mathbf{Z}(\widehat{\omega})^T \mathbf{Y}, \tag{4.27}$$

where $\mathbf{Z}(\omega)$ is defined in (3.5).

The motivation behind using a correction factor, one-fourth of the standard Newton–Raphson correction factor, is based on the following limiting result. As assumed before, ω^0 is the true value of ω.

Theorem 4.5 *Assume that $\widetilde{\omega}$ is an estimate of ω^0 such that $\widetilde{\omega} - \omega^0 = O_p(n^{-1-\delta})$, $\delta \in \left(0, \frac{1}{2}\right]$. Suppose $\widetilde{\omega}$ is updated as $\widehat{\omega}$, using $\widehat{\omega} = \widetilde{\omega} - \frac{1}{4} \times \frac{S'(\widetilde{\omega})}{S''(\widetilde{\omega})}$, then*

(a) $\widehat{\omega} - \omega^0 = O_p(n^{-1-3\delta})$ if $\delta \le \frac{1}{6}$,

(b) $n^{\frac{3}{2}}(\widehat{\omega} - \omega^0) \xrightarrow{d} \mathcal{N}\left(0, \frac{6\sigma^2 c(\omega^0)}{A^{0^2}+B^{0^2}}\right)$ if $\delta > \frac{1}{6}$,

where $c(\omega^0)$ is same as defined in asymptotic distribution of LSEs.

Proof Write

$$\mathbf{D} = \text{diag}\{1, 2, \ldots, n\}, \quad \mathbf{E} = \begin{bmatrix} 0 & 1 \\ -1 & 0 \end{bmatrix}, \tag{4.28}$$

$$\dot{\mathbf{Z}} = \frac{d}{d\omega}\mathbf{Z} = \mathbf{DZE}, \quad \ddot{\mathbf{Z}} = \frac{d^2}{d\omega^2}\mathbf{Z} = -\mathbf{D}^2\mathbf{Z}. \tag{4.29}$$

In this proof, we use $\mathbf{Z}(\omega) \equiv \mathbf{Z}$. Note that $\mathbf{EE} = -\mathbf{I}$, $\mathbf{EE}^T = \mathbf{I} = \mathbf{E}^T\mathbf{E}$ and

$$\frac{d}{d\omega}(\mathbf{Z}^T\mathbf{Z})^{-1} = -(\mathbf{Z}^T\mathbf{Z})^{-1}[\dot{\mathbf{Z}}^T\mathbf{Z} + \mathbf{Z}^T\dot{\mathbf{Z}}](\mathbf{Z}^T\mathbf{Z})^{-1}. \tag{4.30}$$

Compute the first and second derivatives of $S(\omega)$ as;

$$\frac{1}{2}S'(\omega) = \mathbf{Y}^T\dot{\mathbf{Z}}(\mathbf{Z}^T\mathbf{Z})^{-1}\mathbf{Z}^T\mathbf{Y} - \mathbf{Y}^T\mathbf{Z}(\mathbf{Z}^T\mathbf{Z})^{-1}\dot{\mathbf{Z}}^T\mathbf{Z}(\mathbf{Z}^T\mathbf{Z})^{-1}\mathbf{Z}^T\mathbf{Y},$$

$$\frac{1}{2}S''(\omega) = \mathbf{Y}^T\ddot{\mathbf{Z}}(\mathbf{Z}^T\mathbf{Z})^{-1}\mathbf{Z}^T\mathbf{Y} - \mathbf{Y}^T\dot{\mathbf{Z}}(\mathbf{Z}^T\mathbf{Z})^{-1}(\dot{\mathbf{Z}}^T\mathbf{Z} + \mathbf{Z}^T\dot{\mathbf{Z}})(\mathbf{Z}^T\mathbf{Z})^{-1}\mathbf{Z}^T\mathbf{Y}$$

$$+\mathbf{Y}^T\dot{\mathbf{Z}}(\mathbf{Z}^T\mathbf{Z})^{-1}\dot{\mathbf{Z}}^T\mathbf{Y} - \mathbf{Y}^T\dot{\mathbf{Z}}(\mathbf{Z}^T\mathbf{Z})^{-1}\dot{\mathbf{Z}}^T\mathbf{Z}(\mathbf{Z}^T\mathbf{Z})^{-1}\mathbf{Z}^T\mathbf{Y}$$

$$+\mathbf{Y}^T\mathbf{Z}(\mathbf{Z}^T\mathbf{Z})^{-1}(\dot{\mathbf{Z}}^T\mathbf{Z} + \mathbf{Z}^T\dot{\mathbf{Z}})(\mathbf{Z}^T\mathbf{Z})^{-1}\dot{\mathbf{Z}}^T\mathbf{Z}(\mathbf{Z}^T\mathbf{Z})^{-1}\mathbf{Z}^T\mathbf{Y}$$

$$-\mathbf{Y}^T\mathbf{Z}(\mathbf{Z}^T\mathbf{Z})^{-1}(\ddot{\mathbf{Z}}^T\mathbf{Z})(\mathbf{Z}^T\mathbf{Z})^{-1}\mathbf{Z}^T\mathbf{Y}$$

$$-\mathbf{Y}^T\mathbf{Z}(\mathbf{Z}^T\mathbf{Z})^{-1}(\dot{\mathbf{Z}}^T\dot{\mathbf{Z}})(\mathbf{Z}^T\mathbf{Z})^{-1}\mathbf{Z}^T\mathbf{Y}$$

$$+\mathbf{Y}^T\mathbf{Z}(\mathbf{Z}^T\mathbf{Z})^{-1}\dot{\mathbf{Z}}^T\mathbf{Z}(\mathbf{Z}^T\mathbf{Z})^{-1}(\dot{\mathbf{Z}}^T\mathbf{Z} + \mathbf{Z}^T\dot{\mathbf{Z}})(\mathbf{Z}^T\mathbf{Z})^{-1}\mathbf{Z}^T\mathbf{Y}$$

$$-\mathbf{Y}^T\mathbf{Z}(\mathbf{Z}^T\mathbf{Z})^{-1}\dot{\mathbf{Z}}^T\mathbf{Z}(\mathbf{Z}^T\mathbf{Z})^{-1}\dot{\mathbf{Z}}^T\mathbf{Y}.$$

Assume $\tilde{\omega} - \omega^0 = O_p(n^{-1-\delta})$. So, for large n,

$$\frac{1}{n}\mathbf{Z}^T\mathbf{Z}) \equiv (\frac{1}{n}\mathbf{Z}(\tilde{\omega})^T\mathbf{Z}(\tilde{\omega}))^{-1}$$

$$= 2\,\mathbf{I_2} + O_p(\frac{1}{n})$$

and

$$\frac{1}{2n^3}S''(\tilde{\omega}) = \frac{2}{n^4}\mathbf{Y}^T\ddot{\mathbf{Z}}\mathbf{Z}^T\mathbf{Y} - \frac{4}{n^5}\mathbf{Y}^T\dot{\mathbf{Z}}(\dot{\mathbf{Z}}^T\mathbf{Z} + \mathbf{Z}^T\dot{\mathbf{Z}})\mathbf{Z}^T\mathbf{Y} + \frac{2}{n^4}\mathbf{Y}^T\dot{\mathbf{Z}}\dot{\mathbf{Z}}^T\mathbf{Y}$$

$$-\frac{4}{n^5}\mathbf{Y}^T\dot{\mathbf{Z}}\dot{\mathbf{Z}}^T\mathbf{Z}\mathbf{Z}^T\mathbf{Y} + \frac{8}{n^6}\mathbf{Y}^T\mathbf{Z}(\dot{\mathbf{Z}}^T\mathbf{Z} + \mathbf{Z}^T\dot{\mathbf{Z}})\dot{\mathbf{Z}}^T\mathbf{Z}\mathbf{Z}^T\mathbf{Y}$$

$$-\frac{4}{n^5}\mathbf{Y}^T\mathbf{Z}\ddot{\mathbf{Z}}^T\mathbf{Z}\mathbf{Z}^T\mathbf{Y} - \frac{4}{n^5}\mathbf{Y}^T\mathbf{Z}\dot{\mathbf{Z}}^T\dot{\mathbf{Z}}\mathbf{Z}^T\mathbf{Y}$$

$$+\frac{8}{n^6}\mathbf{Y}^T\mathbf{Z}\dot{\mathbf{Z}}^T\mathbf{Z}(\dot{\mathbf{Z}}^T\mathbf{Z} + \mathbf{Z}^T\dot{\mathbf{Z}})\mathbf{Z}^T\mathbf{Y} - \frac{4}{n^5}\mathbf{Y}^T\mathbf{Z}\dot{\mathbf{Z}}^T\mathbf{Z}\dot{\mathbf{Z}}^T\mathbf{Y} + O_p(\frac{1}{n}).$$

Substituting $\dot{\mathbf{Z}}$ and $\ddot{\mathbf{Z}}$ in terms of $\mathbf{D}$ and $\mathbf{Z}$, we obtain

$$
\begin{aligned}
\frac{1}{2n^3}\, S''(\widetilde{\omega}) = {}& -\frac{2}{n^4}\mathbf{Y}^T\mathbf{D}^2\mathbf{Z}\mathbf{Z}^T\mathbf{Y} - \frac{4}{n^5}\mathbf{Y}^T\mathbf{DZE}(\mathbf{E}^T\mathbf{Z}^T\mathbf{DZ}+\mathbf{Z}^T\mathbf{DZE})\mathbf{Z}^T\mathbf{Y} \\
& +\frac{2}{n^4}\mathbf{Y}^T\mathbf{DZEE}^T\mathbf{Z}^T\mathbf{DY} - \frac{4}{n^5}\mathbf{Y}^T\mathbf{DZEE}^T\mathbf{Z}^T\mathbf{DZZ}^T\mathbf{Y} \\
& +\frac{8}{n^6}\mathbf{Y}^T\mathbf{Z}(\mathbf{E}^T\mathbf{Z}^T\mathbf{DZ}+\mathbf{Z}^T\mathbf{DZE})\mathbf{E}^T\mathbf{Z}^T\mathbf{DZZ}^T\mathbf{Y} \\
& +\frac{4}{n^5}\mathbf{Y}^T\mathbf{ZZ}^T\mathbf{D}^2\mathbf{ZZ}^T\mathbf{Y} - \frac{4}{n^5}\mathbf{Y}^T\mathbf{ZE}^T\mathbf{Z}^T\mathbf{D}^2\mathbf{ZEZ}^T\mathbf{Y} \\
& +\frac{8}{n^6}\mathbf{Y}^T\mathbf{ZE}^T\mathbf{Z}^T\mathbf{DZ}(\mathbf{E}^T\mathbf{Z}^T\mathbf{DZ}+\mathbf{Z}^T\mathbf{DZE})\mathbf{Z}^T\mathbf{Y} \\
& -\frac{4}{n^5}\mathbf{Y}^T\mathbf{ZE}^T\mathbf{Z}^T\mathbf{DZE}^T\mathbf{Z}^T\mathbf{DY} + O_p(\frac{1}{n}).
\end{aligned}
$$

Using results (2.5–2.7), one can see that

$$
\frac{1}{n^2}\mathbf{Y}^T\mathbf{DZ} = \frac{1}{4}(A\ B) + O_p(\frac{1}{n}), \quad \frac{1}{n^3}\mathbf{Y}^T\mathbf{D}^2\mathbf{Z} = \frac{1}{6}(A\ B) + O_p(\frac{1}{n}), \quad (4.31)
$$

$$
\frac{1}{n^3}\mathbf{Z}^T\mathbf{D}^2\mathbf{Z} = \frac{1}{6}\mathbf{I}_2 + O_p(\frac{1}{n}), \quad \frac{1}{n}\mathbf{Z}^T\mathbf{Y} = \frac{1}{2}(A\ B)^T + O_p(\frac{1}{n}), \quad (4.32)
$$

$$
\frac{1}{n^2}\mathbf{Z}^T\mathbf{DZ} = \frac{1}{4}\mathbf{I}_2 + O_p(\frac{1}{n}). \quad (4.33)
$$

Therefore,

$$
\begin{aligned}
\frac{1}{2n^3}\, S''(\widetilde{\omega}) &= (A^2 + B^2)\left[-\frac{1}{6} - 0 + \frac{1}{8} - \frac{1}{8} + 0 + \frac{1}{6} - \frac{1}{6} + 0 + \frac{1}{8}\right] + O_p(\frac{1}{n}) \\
&= -\frac{1}{24}(A^2 + B^2) + O_p(\frac{1}{n}).
\end{aligned}
$$

Write $S'(\omega)/2n^3 = I_1 + I_2$ and simplify I_1 and I_2 separately for large n.

$$
\begin{aligned}
I_1 &= \frac{1}{n^3}\mathbf{Y}^T\dot{\mathbf{Z}}(\mathbf{Z}^T\mathbf{Z})^{-1}\mathbf{Z}^T\mathbf{Y} = \frac{2}{n^4}\mathbf{Y}^T\mathbf{DZEZ}^T\mathbf{Y}, \\
I_2 &= \frac{1}{n^3}\mathbf{Y}^T\mathbf{Z}(\mathbf{Z}^T\mathbf{Z})^{-1}\dot{\mathbf{Z}}^T\mathbf{Z}(\mathbf{Z}^T\mathbf{Z})^{-1}\mathbf{Z}^T\mathbf{Y} \\
&= \frac{1}{n^3}\mathbf{Y}^T\mathbf{Z}(\mathbf{Z}^T\mathbf{Z})^{-1}\mathbf{E}^T\mathbf{Z}^T\mathbf{DZ}(\mathbf{Z}^T\mathbf{Z})^{-1}\mathbf{Z}^T\mathbf{Y} \\
&= \frac{1}{n^3}\mathbf{Y}^T\mathbf{Z}(2\mathbf{I} + O_p(\frac{1}{n}))\mathbf{E}^T(\frac{1}{4}\mathbf{I} + O_p(\frac{1}{n}))(2\mathbf{I} + O_p(\frac{1}{n}))\mathbf{Z}^T\mathbf{Y} \\
&= \frac{1}{n^3}\mathbf{Y}^T\mathbf{ZE}^T\mathbf{Z}^T\mathbf{Y} + O_p(\frac{1}{n}) = O_p(\frac{1}{n}),
\end{aligned}
$$

and $(n^4 I_1)/2 = \mathbf{Y}^T \mathbf{DZEZ}^T \mathbf{Y}$ at $\widetilde{\omega}$ for large n is simplified as

$$
\mathbf{Y}^T \mathbf{DZEZ}^T \mathbf{Y} = \mathbf{Y}^T \mathbf{DZ}(\widetilde{\omega})\mathbf{EZ}^T(\widetilde{\omega})\mathbf{Y}
$$

$$
= \left(\sum_{t=1}^{n} y(t)t\cos(\widetilde{\omega}t)\right)\left(\sum_{t=1}^{n} y(t)\sin(\widetilde{\omega}t)\right)
$$

$$
- \left(\sum_{t=1}^{n} y(t)t\sin(\widetilde{\omega}t)\right)\left(\sum_{t=1}^{n} y(t)\cos(\widetilde{\omega}t)\right).
$$

Observe that $\displaystyle\sum_{t=1}^{n} y(t)e^{-i\omega t} = \sum_{t=1}^{n} y(t)\cos(\omega t) - i\sum_{t=1}^{n} y(t)\sin(\omega)$. Then along the same line as Bai et al. [19] (see also Nandi and Kundu [20])

$$
\sum_{t=1}^{n} y(t)e^{-i\widetilde{\omega}t} = \sum_{t=1}^{n}\left[A^0\cos(\omega^0 t) + B^0\sin(\omega^0 t) + X(t)\right]e^{-i\widetilde{\omega}t}
$$

$$
= \sum_{t=1}^{n}\left[\frac{A^0}{2}\left(e^{i\omega^0 t} + e^{-i\omega^0 t}\right) + \frac{B^0}{2i}\left(e^{i\omega^0 t} - e^{-i\omega^0 t}\right) + X(t)\right]e^{-i\widetilde{\omega}t}
$$

$$
= \left(\frac{A^0}{2} + \frac{B^0}{2i}\right)\sum_{t=1}^{n} e^{i(\omega^0 - \widetilde{\omega})t} + \left(\frac{A^0}{2} - \frac{B^0}{2i}\right)\sum_{t=1}^{n} e^{-i(\omega^0 + \widetilde{\omega})t}
$$

$$
+ \sum_{t=1}^{n} X(t)e^{-i\widetilde{\omega}t}.
$$

If $\widetilde{\omega} - \omega^0 = O_p(n^{-1-\delta})$, the it can be shown that $\displaystyle\sum_{t=1}^{n} e^{-i(\omega^0 + \widetilde{\omega})t} = O_p(1)$ and

$$
\sum_{t=1}^{n} e^{i(\omega^0 - \widetilde{\omega})t} = n + i(\omega^0 - \widetilde{\omega})\sum_{t=1}^{n} e^{i(\omega^0 - \omega^*)t} = n + O_p(n^{-1-\delta})O_p(n^2)
$$

$$
= n + O_p(n^{1-\delta}),
$$

where ω^* is a point between ω^0 and $\widetilde{\omega}$. Choose L_1 large enough, such that $L_1\delta > 1$. Therefore, using Taylor series approximation of $e^{-i\widetilde{\omega}_j t}$ around ω^0 up to L_1th order terms

$$
\sum_{t=1}^{n} X(t)e^{-i\widetilde{\omega}t}
$$

$$
= \sum_{k=0}^{\infty} a(k)\sum_{t=1}^{n} e(t-k)e^{-i\widetilde{\omega}t}
$$

$$= \sum_{k=0}^{\infty} a(k) \sum_{t=1}^{n} e(t-k)e^{-i\omega^0 t} + \sum_{k=0}^{\infty} a(k) \sum_{l=1}^{L_1-1} \frac{(-i(\tilde{\omega}-\omega^0))^l}{l!} \sum_{t=1}^{n} e(t-k)t^l e^{-i\omega^0 t}$$

$$+ \sum_{k=0}^{\infty} a(k) \frac{\theta_1(n(\tilde{\omega}-\omega^0))^{L_1}}{L_1!} \sum_{t=1}^{n} |e(t-k)|,$$

with $|\theta_1| < 1$. Since $\{a(k)\}$ is absolutely summable, $\sum_{k=0}^{\infty} |a(k)| < \infty$,

$$\sum_{t=1}^{n} X(t)e^{-i\tilde{\omega}t} = O_p(n^{\frac{1}{2}}) + \sum_{l=1}^{L_1-1} \frac{O_p(n^{-(1+\delta)l})}{l!} O_p(n^{l+\frac{1}{2}}) + O_p\left((n.n^{-1-\delta})^{L_1}.n\right)$$

$$= O_p(n^{\frac{1}{2}}) + O_p(n^{\frac{1}{2}+\delta-L_1\delta}) + O_p(n^{1-L_1\delta}) = O_p(n^{\frac{1}{2}}).$$

Therefore,

$$\sum_{t=1}^{n} y(t)e^{-i\tilde{\omega}t} = \left(\frac{A^0}{2} + \frac{B^0}{2i}\right)\left(n + O_p(n^{1-\delta})\right) + O_p(1) + O_p(n^{\frac{1}{2}})$$

$$= \frac{n}{2}\left[(A^0 - iB^0) + O_p(n^{-\delta})\right] \quad \text{as} \quad \delta \in \left(0, \frac{1}{2}\right],$$

and

$$\sum_{t=1}^{n} y(t)\cos(\tilde{\omega}t) = \frac{n}{2}\left(A^0 + O_p(n^{-\delta})\right), \quad \sum_{t=1}^{n} y(t)\sin(\tilde{\omega}t) = \frac{n}{2}\left(B^0 + O_p(n^{-\delta})\right).$$

Similarly as above, observe that

$$\sum_{t=1}^{n} y(t)te^{-i\tilde{\omega}t} = \sum_{t=1}^{n}\left(A^0\cos(\omega^0 t) + B^0\sin(\omega^0 t) + X(t)\right)te^{-i\tilde{\omega}t}$$

$$= \frac{1}{2}(A^0 - iB^0)\sum_{t=1}^{n} t\, e^{i(\omega^0-\tilde{\omega})t}$$

$$+ \frac{1}{2}(A^0 + iB^0)\sum_{t=1}^{n} t\, e^{-i(\omega^0+\tilde{\omega})t} + \sum_{t=1}^{n} X(t)te^{-i\tilde{\omega}t},$$

and following Bai et al. [19]

$$\sum_{t=1}^{n} t\, e^{-i(\omega^0 + \widetilde{\omega})t} = O_p(n),$$

$$\sum_{t=1}^{n} t\, e^{i(\omega^0 - \widetilde{\omega})t} = \sum_{t=1}^{n} t + i(\omega^0 - \widetilde{\omega})\sum_{t=1}^{n} t^2 - \frac{1}{2}(\omega^0 - \widetilde{\omega})^2 \sum_{t=1}^{n} t^3$$

$$- \frac{1}{6}i(\omega^0 - \widetilde{\omega})^3 \sum_{t=1}^{n} t^4 + \frac{1}{24}(\omega^0 - \widetilde{\omega})^4 \sum_{t=1}^{n} t^5 e^{i(\omega^0 - \omega^*)t}.$$

Again using $\widetilde{\omega} - \omega^0 = O_p(n^{-1-\delta})$, we have

$$\frac{1}{24}(\omega^0 - \widetilde{\omega})^4 \sum_{t=1}^{n} t^5 e^{i(\omega^0 - \omega^*)t} = O_p(n^{2-4\delta}). \tag{4.34}$$

Choose L_2 large enough such that $L_2\delta > 1$ and using Taylor series expansion of $e^{-i\widetilde{\omega}t}$ we have,

$$\sum_{t=1}^{n} X(t)t e^{-i\widetilde{\omega}t} = \sum_{k=0}^{\infty} a(k) \sum_{t=1}^{n} e(t-k)t e^{-i\widetilde{\omega}t}$$

$$= \sum_{k=0}^{\infty} a(k) \sum_{t=1}^{n} e(t-k)t e^{-i\omega^0 t}$$

$$+ \sum_{k=0}^{\infty} a(k) \sum_{l=1}^{L_2-1} \frac{(-i(\widetilde{\omega} - \omega^0))^l}{l!} \sum_{t=1}^{n} e(t-k)t^{l+1} e^{-i\omega^0 t}$$

$$+ \sum_{k=0}^{\infty} a(k) \frac{\theta_2 (n(\widetilde{\omega} - \omega^0))^{L_2}}{L_2!} \sum_{t=1}^{n} t|e(t-k)| \ (\text{as before } |\theta_2| < 1)$$

$$= \sum_{k=0}^{\infty} a(k) \sum_{t=1}^{n} e(t-k)t e^{-i\omega^0 t} + \sum_{l=1}^{L_2-1} O_p(n^{-(1+\delta)l}) O_p(n^{l+\frac{3}{2}})$$

$$+ \sum_{k=0}^{\infty} a(k) O_p(n^{\frac{5}{2} - L_2\delta})$$

$$= \sum_{k=0}^{\infty} a(k) \sum_{t=1}^{n} e(t-k)t e^{-i\omega^0 t} + O_p(n^{\frac{5}{2} - L_2\delta}).$$

Therefore,

$$\sum_{t=1}^{n} y(t)t\cos(\widetilde{\omega}t)$$

$$= \frac{1}{2}\left[A^0\left(\sum_{t=1}^{n}t - \frac{1}{2}(\omega^0-\widetilde{\omega})^2\sum_{t=1}^{n}t^3\right) + B^0\left(\sum_{t=1}^{n}(\omega^0-\widetilde{\omega})t^2 - \frac{1}{6}(\omega^0-\widetilde{\omega})^3\sum_{t=1}^{n}t^4\right)\right]$$

$$+ \sum_{k=0}^{\infty}a(k)\sum_{t=1}^{n}e(t-k)t\cos(\omega^0 t) + O_p(n^{\frac{5}{2}-L_2\delta}) + O_p(n) + O_p(n^{2-4\delta}).$$

Similarly,

$$\sum_{t=1}^{n} y(t)t\sin(\widetilde{\omega}t)$$

$$= \frac{1}{2}\left[B^0\left(\sum_{t=1}^{n}t - \frac{1}{2}(\omega^0-\widetilde{\omega})^2\sum_{t=1}^{n}t^3\right) - A^0\left(\sum_{t=1}^{n}(\omega^0-\widetilde{\omega})t^2 - \frac{1}{6}(\omega^0-\widetilde{\omega})^3\sum_{t=1}^{n}t^4\right)\right]$$

$$+ \sum_{k=0}^{\infty}a(k)\sum_{t=1}^{n}e(t-k)t\sin(\omega^0 t) + O_p(n^{\frac{5}{2}-L_2\delta}) + O_p(n) + O_p(n^{2-4\delta}).$$

Hence,

$$\widehat{\omega} = \widetilde{\omega} - \frac{1}{4}\frac{S'(\widetilde{\omega})}{S''(\widetilde{\omega})}$$

$$= \widetilde{\omega} - \frac{1}{4}\frac{\frac{1}{2n^3}S'(\widetilde{\omega})}{-\frac{1}{24}(A^{0^2}+B^{0^2}) + O_p(\frac{1}{n})}$$

$$= \widetilde{\omega} - \frac{1}{4}\frac{\frac{2}{n^4}\mathbf{Y}^T\mathbf{DZEZ}^T\mathbf{Y}}{-\frac{1}{24}(A^{0^2}+B^{0^2}) + O_p(\frac{1}{n})}$$

$$= \widetilde{\omega} + 12\frac{\frac{1}{n^4}\mathbf{Y}^T\mathbf{DZEZ}^T\mathbf{Y}}{(A^{0^2}+B^{0^2}) + O_p(\frac{1}{n})}$$

$$= \widetilde{\omega} + 12\frac{\frac{1}{4n^3}(A^{0^2}+B^{0^2})\left\{(\omega^0-\widetilde{\omega})\sum_{t=1}^{n}t^2 - \frac{1}{6}(\omega^0-\widetilde{\omega})^3\sum_{t=1}^{n}t^4\right\}}{(A^{0^2}+B^{0^2}) + O_p(\frac{1}{n})}$$

$$+ \left[B^0\sum_{k=0}^{\infty}a(k)\sum_{t=1}^{n}e(t-k)t\cos(\omega^0 t) + A^0\sum_{k=0}^{\infty}a(k)\sum_{t=1}^{n}e(t-k)t\sin(\omega^0 t)\right]$$

$$\times \frac{6}{(A^{0^2}+B^{0^2})n^3 + O_p(\frac{1}{n})} + O_p(n^{-\frac{1}{2}-L_2\delta}) + O_p(n^{-2}) + O_p(n^{-1-4\delta})$$

$$= \omega^0 + (\omega^0-\widetilde{\omega})O_p(n^{-2\delta})$$

$$+ \left[B^0 \sum_{k=0}^{\infty} a(k) \sum_{t=1}^{n} e(t-k)t \cos(\omega^0 t) + A^0 \sum_{k=0}^{\infty} a(k) \sum_{t=1}^{n} e(t-k)t \sin(\omega^0 t) \right]$$

$$\times \frac{6}{(A^{0^2} + B^{0^2})n^3 + O_p(\frac{1}{n})} + O_p(n^{-\frac{1}{2}-L_2\delta}) + O_p(n^{-2}) + O_p(n^{-1-4\delta}).$$

Finally, if $\delta \leq 1/6$, clearly $\widehat{\omega} - \omega^0 = O_p(n^{-1-3\delta})$, and if $\delta > 1/6$, then

$$n^{\frac{3}{2}}(\widehat{\omega} - \omega^0) \stackrel{d}{=} \frac{6n^{-\frac{3}{2}}}{(A^{0^2} + B^{0^2})} \left[B^0 \sum_{k=0}^{\infty} a(k) \sum_{t=1}^{n} e(t-k)t \cos(\omega^0 t) \right.$$

$$\left. + A^0 \sum_{k=0}^{\infty} a(k) \sum_{t=1}^{n} e(t-k)t \sin(\omega^0 t) \right]$$

$$= \frac{6n^{-\frac{3}{2}}}{(A^{0^2} + B^{0^2})} \left[B^0 \sum_{t=1}^{n} X(t)t \cos(\omega^0 t) + A^0 \sum_{t=1}^{n} X(t)t \sin(\omega^0 t) \right]$$

$$\left[\text{Using similar technique described in Appendix B.} \right]$$

$$\stackrel{d}{\longrightarrow} \mathcal{N}\left(0, \frac{6\sigma^2 c(\omega^0)}{A^{0^2} + B^{0^2}} \right).$$

That proves the theorem. $\square$

Remark 4.3 In Eq. (4.27), write $(\widehat{A} \ \widehat{B})^T = (A(\widehat{\omega})B(\widehat{\omega}))^T$. Expanding $A(\widehat{\omega})$ around ω^0 by Taylor series,

$$A(\widehat{\omega}) - A(\omega^0) = (\widehat{\omega} - \omega^0)A'(\bar{\omega}) + o(n^2). \qquad (4.35)$$

$A'(\bar{\omega})$ is the first-order derivative of $A(\omega)$ with respect to ω at $\bar{\omega}$; $\bar{\omega}$ is a point on the line joining $\widehat{\omega}$ and ω^0; $\widehat{\omega}$ can be either the LSE or the estimator obtained by modified Newton-Raphson method. Comparing the variances (asymptotic) of the two estimators of ω, note that the asymptotic variance of the corresponding estimator of A^0 is four times less than that of the LSE. The same is true for the estimator of B^0.

$$\text{Var}(A(\widehat{\omega}) - A(\omega^0)) \approx \text{Var}(\widehat{\omega} - \omega^0)[A'(\bar{\omega})]^2$$

$$\text{Var}(A(\widehat{\omega}) - A^0) = \frac{\text{Var}(\widehat{\omega}_{LSE})}{4}[A'(\bar{\omega})]^2 = \frac{\text{Var}(\widehat{A}_{LSE})}{4},$$

where $\widehat{\omega}_{LSE}$ and $\widehat{A}_{LSE}$ are the LSEs of ω^0 and A^0, respectively. Similarly, $\text{Var}(B(\widehat{\omega}) - B(\omega^0)) = \text{Var}(\widehat{B}_{LSE})/4$ and different notation involving B have a similar meaning replacing A by B.

Remark 4.4 Theorem 4.5 suggests that an initial estimator of ω^0 having convergence rate equal to $O_p(n^{-1-\delta})$ is required to use it. But such an estimator is not easily available. Therefore, Kundu et al. [16] use an estimator having convergence rate $O_p(n^{-1})$ and using a fraction of the available data points to implement Theorem 4.5. The subset is chosen in such a way that the dependence structure of the error process is not destroyed. The argument maximum of $I(\omega)$, defined in (1.5), or $S(\omega)$, defined in (3.55), over Fourier frequencies, provides an estimator of the frequency with convergence rate $O_p(n^{-1})$.

4.6 Multiple Sinusoidal Model

The multi-component frequency model, generally known as multiple sinusoidal model, has the following form in the presence of p distinct frequency component;

$$y(t) = \sum_{k=1}^{p} \left[A_k^0 \cos(\omega_k^0 t) + B_k^0 \sin(\omega_k^0 t) \right] + X(t), \quad t = 1, \ldots, n. \qquad (4.36)$$

Here $\{y(1), \ldots, y(n)\}$ is the observed data. For $k = 1, \ldots, p$, A_k^0 and B_k^0 are amplitudes and ω_k^0 is the frequency. Similar to single-component model (4.1), the additive error $\{X(t)\}$ is the sequence of random variables which satisfies different assumptions depending upon the problem at hand. In this chapter, p is assumed to be known. The problem of estimation of p is considered in Chap. 5.

Extensive work, on multiple sinusoidal signal model, has been done by several authors. Some references: Kundu and Mitra [12] studied the multiple sinusoidal model (4.36) and established the strong consistency and asymptotic normality of the LSEs under Assumption 3.1, that is, $\{X(t)\}$ is sequence of i.i.d. random variables with finite second moment. Kundu [10] proved the same results under Assumption 3.2. Irizarry [13] consider a semi-parametric weighted LSEs of the parameters of model (4.36) and develop the asymptotic variance expression (discussed in next section).

The LSEs of the unknown parameters are asymptotically normality distributed under Assumption 3.2 and stated in the following theorem.

Theorem 4.6 *Under Assumption 3.2, as $n \to \infty$, $\{n^{\frac{1}{2}}(\widehat{A}_k - A_k^0), n^{\frac{1}{2}}(\widehat{B}_k - B_k^0),$ $n^{\frac{3}{2}}(\widehat{\omega}_k - \omega_k^0)\}$ are jointly distributed as a 3-variate normal distribution with mean vector zero and dispersion matrix $2\,\sigma^2 c(\omega_k^0)\,\boldsymbol{\Sigma}_k^{-1}$, where $c(\omega_k^0)$ and $\boldsymbol{\Sigma}_k^{-1}$ are same as $c(\omega^0)$ and $\boldsymbol{\Sigma}^{-1}$, with A^0, B^0, and ω^0 replaced by A_k^0, B_k^0, and ω_k^0, respectively. For $j \neq k$, $(\{n^{\frac{1}{2}}(\widehat{A}_j - A_j^0), n^{\frac{1}{2}}(\widehat{B}_j - B_j^0), n^{\frac{3}{2}}(\widehat{\omega}_j - \omega_j^0)\})$, and $(\{n^{\frac{1}{2}}(\widehat{A}_k - A_k^0), n^{\frac{1}{2}}(\widehat{B}_k - B_k^0), n^{\frac{3}{2}}(\widehat{\omega}_k - \omega_k^0)\})$ are asymptotically independently distributed. The quantities $c(\omega^0)$ and $\boldsymbol{\Sigma}^{-1}$ are defined in (4.10) and (4.11), respectively.*

4.7 Weighted Least Squares Estimators

The WLSEs are proposed by Irizarry [13] in order to analyze the harmonic components in musical sounds. The least squares results in case of multiple sinusoidal model are extended to weighted least squares case for a class of weight functions. Under the assumption that the error process is stationary with certain other properties, the asymptotic results are developed. The WLSEs minimize the following criterion function

$$
S(\boldsymbol{\omega}, \boldsymbol{\eta}) = \sum_{t=1}^{n} w\left(\frac{t}{n}\right)\left(y(t) - \sum_{k=1}^{p}(A_k \cos(\omega_k t) + B_k \sin(\omega_k t))\right)^2, \qquad (4.37)
$$

with respect to $\boldsymbol{\omega} = (\omega_1, \ldots, \omega_p)$ and $\boldsymbol{\eta} = (A_1, \ldots A_p, B_1, \ldots, B_p)$. The error process $\{X(t)\}$ is stationary with autocovariance function $c_{xx}(h) = Cov(X(t), X(t+h))$ and spectral density function $f_x(\lambda)$, $-\infty < \lambda < \infty$ satisfies Assumption 4.3 and the weight function $w(s)$ satisfies Assumption 4.4.

Assumption 4.3 The error process $\{X(t)\}$ is a strictly stationary real-valued random process all of whose moments exist, with zero mean and with $c_{xx...x}(h_1, \ldots, h_{L-1})$, the joint cumulant function of order L for $L = 2, 3, \ldots$. Also,

$$
C_L = \sum_{h_1=-\infty}^{\infty} \cdots \sum_{h_{L-1}=-\infty}^{\infty} |c_{xx...x}(h_1, \ldots, h_{L-1})|
$$

satisfy $\sum_{k}(C_k z^k)/k! < \infty$, for z in a neighborhood of 0.

Assumption 4.4 The weight function $w(s)$ is non-negative, bounded, of bounded variation, has support [0, 1] and it is such that $W_0 > 0$ and $W_1^2 - W_0 W_2 \neq 0$, where

$$
W_n = \int_0^1 s^n w(s) ds. \qquad (4.38)
$$

Irrizarry [13] uses the idea of Walker [7] discussed for the unweighted case by proving the following Lemma.

Lemma 4.5 *If $w(t)$ satisfies Assumption 4.4, then for $k = 0, 1, \ldots$*

$$
\lim_{n \to \infty} \frac{1}{n^{k+1}} \sum_{t=1}^{n} w\left(\frac{t}{n}\right) t^k \exp(i\lambda t) = W_n, \qquad for \ \lambda = 0, 2\pi,
$$

$$
\sum_{t=1}^{n} w\left(\frac{t}{n}\right) t^k \exp(i\lambda t) = O(n^k), \qquad for \ 0 < \lambda < 2\pi.
$$

Similar to the unweighted case, that is the LSEs, the WLSEs of (ω, η) are asymptotically equivalent to the weighted PEs which maximize

$$I_{Wy}(\omega) = \sum_{k=1}^{p} \left| \frac{1}{n} \sum_{t=1}^{n} w\left(\frac{t}{n}\right) y(t) \exp(it\omega_k) \right|^2$$

with respect to ω and the estimators of η are given in Sect. 3.15. This is the sum of the periodogram functions of the tapered data $w\left(t/n\right) y(t)$. The following lemma is used to prove the consistency and asymptotic normality of the WLSEs.

Lemma 4.6 *Let the error process* $\{X(t)\}$ *satisfy Assumption 4.3 and the weight function* $w(s)$ *satisfy Assumption 4.4, then*

$$\lim_{n\to\infty} \sum_{0\leq\lambda\leq 2\pi} \left| \frac{1}{n^{k+1}} \sum_{t=n}^{\infty} w\left(\frac{t}{n}\right) t^k X(t) \exp(-it\lambda) \right| = 0, \quad \textit{in probability.}$$

The consistency and asymptotic normality of the WLSEs are stated in Theorem 4.7 and Theorem 4.8, respectively. Let $\omega^0 = (\omega_1^0, \ldots \omega_p^0)$ and $\eta = (A_1^0, \ldots A_p^0, B_1^0, \ldots, B_p^0)$ be the true parameter values.

Theorem 4.7 *Under Assumptions 4.3 and 4.4, for* $0 < \omega_k^0 < \pi$,

$$\widehat{A}_k \xrightarrow{p} A_k^0, \quad \textit{and} \quad \widehat{B}_k \xrightarrow{p} B_k^0, \quad \textit{as } n \to \infty$$

$$\lim_{n\to\infty} n|\widehat{\omega}_k - \omega_k^0| = 0, \quad \textit{for } k = 1, \ldots, p.$$

Theorem 4.8 *Under the same assumptions as in Theorem 4.7,*

$$\left(n^{\frac{1}{2}}(\widehat{A}_k - A_k^0), n^{\frac{1}{2}}(\widehat{B}_k - B_k^0), n^{\frac{3}{2}}(\widehat{\omega}_k - \omega_k^0) \right)' \xrightarrow{d} \mathscr{N}_3\left(0, \frac{4\pi f_x(\omega_k^0)}{A_k^{0^2} + B_k^{0^2}} \mathbf{V_k} \right)$$

where

$$\mathbf{V_k} = \begin{bmatrix} c_1 A_k^{0^2} + c_2 B_k^{0^2} & -c_3 A_k^0 B_k^0 & -c_4 B_k^0 \\ -c_3 A_k^0 B_k^0 & c_2 A_k^{0^2} + c_1 B_k^{0^2} & c_4 A_k^0 \\ -c_4 B_k^0 & c_4 A_k^0 & c_0 \end{bmatrix}$$

with

$$c_0 = a_0 b_0$$
$$c_1 = U_0 W_0^{-1}$$
$$c_2 = a_0 b_1$$
$$c_3 = a_0 W_1 W_0^{-1} (W_o^2 W_1 U_2 - W_1^3 U_0 - 2W_0^2 + 2W_0 W_1 W_2 U_0)$$
$$c_4 = a_0 (W_0 W_1 U_2 - W_1^2 U_1 - W_0 W_2 U_1 + W_1 W_2 U_0),$$

and

$$a_0 = (W_0 W_2 - W_1^2)^{-2}$$
$$b_i = W_i^2 U_2 + W_{i+1}(W_{i+1} U_0 - 2W_i U_1), \quad for \ \ i = 0, 1.$$

Here, W_i, $i = 0, 1, 2$ are defined in (4.38) and U_i, $i = 0, 1, 2$ are defined by

$$U_i = \int_0^1 s^i w(s)^2 ds.$$

4.8 Conclusions

In this chapter, we mainly emphasize the asymptotic properties of LSEs of the unknown parameters of sinusoidal signal model under different error assumptions. The theoretical properties of the superefficient estimator are discussed in much detail because it has the same rate of convergence as the LSEs. At the same time it has a smaller asymptotic variance than the LSEs. Some of the other estimation procedures, presented in Chap. 3, have desirable theoretical properties. Bai et al. [21] proved the consistency of EVLP estimators, whereas Kundu and Mitra [22] did the same for NSD estimators. The proofs of convergence of the modified Prony algorithm and constrained MLE are found in Kundu [23] and Kundu and Kannan [24], respectively. Sequential estimators are strongly consistent and have the same limiting distribution as the LSEs, see Prasad et al. [15]. The frequency estimator obtained by using Quinn and Fernandes method is strongly consistent and as efficient as the LSEs, Quinn and Fernandes [9]. Trung-Van [25] proved that the estimators of the frequencies obtained by amplitude harmonics method are strongly consistent, their bias converges to zero almost surely with rate $n^{-3/2}(\log n)^\delta$, $\delta > 1/2$ and have the same asymptotic variances as Whittle estimators. Nandi and Kundu [20] proved that the algorithm presented in Sect. 3.16 is as efficient as the LSEs.

Appendix A

Proof of Lemma 4.1

In this proof, we denote $\widehat{\boldsymbol{\theta}}$ by $\widehat{\boldsymbol{\theta}}_n = (\widehat{A}_n, \widehat{B}_n, \widehat{\omega}_n)$ and $Q(\boldsymbol{\theta})$ by $Q_n(\boldsymbol{\theta})$ to emphasize that they depend on n. Suppose (4.3) is true and $\widehat{\boldsymbol{\theta}}_n$ does not converge to $\boldsymbol{\theta}^0$ as $n \to \infty$, then there exist a $c > 0$, an $0 < M < \infty$ and a subsequence $\{n_k\}$ of $\{n\}$, such that $\widehat{\boldsymbol{\theta}}_{n_k} \in S_{c,M}$ for all $k = 1, 2, \ldots$. Since $\widehat{\boldsymbol{\theta}}_{n_k}$ is the LSE of $\boldsymbol{\theta}^0$ when $n = n_k$,

$$Q_{n_k}(\widehat{\boldsymbol{\theta}}_{n_k}) \leq Q_{n_k}(\boldsymbol{\theta}^0) \Rightarrow \frac{1}{n_k}\left[Q_{n_k}(\widehat{\boldsymbol{\theta}}_{n_k}) - Q_{n_k}(\boldsymbol{\theta}^0)\right] \leq 0.$$

Therefore,

$$\varliminf_{\boldsymbol{\theta}_{n_k} \in S_{c,M}} \frac{1}{n_k}[Q_{n_k}(\widehat{\boldsymbol{\theta}}_{n_k}) - Q_{n_k}(\boldsymbol{\theta}^0)] \leq 0,$$

which contradicts the inequality (4.3). Thus, $\widehat{\boldsymbol{\theta}}_n$ is a strongly consistent estimator of $\boldsymbol{\theta}^0$. $\qquad\square$

Proof of Lemma 4.2 under Assumption 4.1

We prove the result for $\cos(\omega t)$, the result for $\sin(\omega t)$ follows similarly. Let $Z(t) = X(t)I_{[|X(t)|\leq t^{\frac{1}{1+\delta}}]}$. Then

$$\sum_{t=1}^{\infty} P[Z(t) \neq X(t)] = \sum_{t=1}^{\infty} P[|X(t)| > t^{\frac{1}{1+\delta}}] = \sum_{t=1}^{\infty} \sum_{2^{t-1}\leq m<2^t} P[|X(1)| > m^{\frac{1}{1+\delta}}]$$

$$\leq \sum_{t=1}^{\infty} 2^t P[2^{\frac{t-1}{1+\delta}} \leq |X(1)|]$$

$$\leq \sum_{t=1}^{\infty} 2^t \sum_{k=t}^{\infty} P[2^{\frac{k-1}{1+\delta}} \leq |X(1)| < 2^{\frac{k}{1+\delta}}]$$

$$\leq \sum_{k=1}^{\infty} P[2^{\frac{k-1}{1+\delta}} \leq |X(1)| < 2^{\frac{k}{1+\delta}}] \sum_{t=1}^{k} 2^t$$

$$\leq C \sum_{k=1}^{\infty} 2^{k-1} P[2^{\frac{k-1}{1+\delta}} \leq |X(1)| < 2^{\frac{k}{1+\delta}}]$$

$$\leq C \sum_{k=1}^{\infty} E|X(1)|^{1+\delta} I_{[2^{\frac{k-1}{1+\delta}}\leq|X(1)|<2^{\frac{k}{1+\delta}}]} \leq CE|X(1)|^{1+\delta} < \infty.$$

Therefore, $P[Z(t) \neq X(t) \text{ i.o.}] = 0$. Thus

$$\sup_{0 \leq \omega \leq 2\pi} \frac{1}{n} \sum_{t=1}^{n} X(t) \cos(\omega t) \to 0 \text{ a.s} \Leftrightarrow \sup_{0 \leq \omega \leq 2\pi} \frac{1}{n} \sum_{t=1}^{n} Z(t) \cos(\omega t) \to 0 \text{ a.s.}$$

Let $U(t) = Z(t) - E(Z(t))$, then

$$\sup_{0 \leq \omega \leq 2\pi} \left| \frac{1}{n} \sum_{t=1}^{n} E(Z(t)) \cos(\omega t) \right| \leq \frac{1}{n} \sum_{t=1}^{n} |E(Z(t))|$$

$$= \frac{1}{n} \sum_{t=1}^{n} \left| \int_{-t^{\frac{1}{1+\delta}}}^{t^{\frac{1}{1+\delta}}} x \, dF(x) \right| \to 0.$$

Thus, we only need to show that

$$\sup_{0 \leq \omega \leq 2\pi} \frac{1}{n} \sum_{t=1}^{n} U(t) \cos(\omega t) \to 0 \quad \text{a.s.}$$

For any fixed ω and $\varepsilon > 0$, let $0 \leq h \leq \frac{1}{2n^{\frac{1}{1+\delta}}}$, then we have

$$P\left\{ \left| \frac{1}{n} \sum_{t=1}^{n} U(t) \cos(\omega t) \right| \geq \varepsilon \right\} \leq 2e^{-hn\varepsilon} \prod_{t=1}^{n} E e^{hU(t)\cos(\omega t)}$$

$$\leq 2e^{-hn\varepsilon} \prod_{t=1}^{n} \left(1 + 2Ch^{1+\delta} \right).$$

Since $|hU(t)\cos(\omega t)| \leq \frac{1}{2}$, $e^x \leq 1 + x + 2|x|^{1+\delta}$ for $|x| \leq \frac{1}{2}$ and $E|U(t)|^{1+\delta} < C$ for some $C > 0$. Clearly,

$$2e^{-hn\varepsilon} \prod_{t=1}^{n} \left(1 + 2Ch^{1+\delta} \right) \leq 2e^{-hn\varepsilon + 2nCh^{1+\delta}}.$$

Choose $h = 1/(2n^{-(1+\delta)}$, then for large n,

$$P\left\{ \left| \frac{1}{n} \sum_{t=1}^{n} U(t) \cos(\omega t) \right| \geq \varepsilon \right\} \leq 2e^{-\frac{\varepsilon}{2}n^{\frac{\delta}{1+\delta}} + C} \leq Ce^{-\frac{\varepsilon}{2}n^{\frac{\delta}{1+\delta}}}.$$

Let $K = n^2$, choose $\omega_1, \ldots, \omega_K$, such that for each $\omega \in (0, 2\pi)$, we have a ω_k satisfying $|\omega_k - \omega| \leq 2\pi/n^2$. Note that

$$\left| \frac{1}{n} \sum_{t=1}^{n} U(t) \left(\cos(\omega t) - \cos(\omega_k t) \right) \right| \leq C \frac{1}{n} \sum_{t=1}^{n} t^{\frac{1}{1+\delta}} . t . \left(\frac{2\pi}{n^2} \right) \leq C \pi n^{-\frac{\delta}{1+\delta}} \to 0.$$

Therefore for large n, we have

$$P \left\{ \sup_{0 \leq \omega \leq 2\pi} \left| \frac{1}{n} \sum_{t=1}^{n} U(t) \cos(\omega t) \right| \geq 2\varepsilon \right\}$$

$$\leq P \left\{ \max_{k \leq n^2} \left| \frac{1}{n} \sum_{t=1}^{n} U(t) \cos(\omega t_k) \right| \geq \varepsilon \right\} \leq C n^2 e^{-\frac{\varepsilon}{2} n^{\frac{\delta}{1+\delta}}}.$$

Since $\sum_{n=1}^{\infty} n^2 e^{-\frac{\varepsilon}{2} n^{\frac{\delta}{1+\delta}}} < \infty$, therefore

$$\sup_{0 \leq \omega \leq 2\pi} \left| \frac{1}{n} \sum_{t=1}^{n} U(t) \cos(\omega t) \right| \to 0 \quad \text{a.s.}$$

by Borel Cantelli lemma.   □

Proof of Lemma 4.2 under Assumption 3.2

Under Assumption 3.2, the error process $\{X(t)\}$ is a stationary linear process with absolutely summable coefficients. Observe that (Kundu [10])

$$\frac{1}{n} \sum_{t=1}^{n} X(t) \cos(\omega t) = \frac{1}{n} \sum_{t=1}^{n} \sum_{j=0}^{\infty} a(j) e(t - j) \cos(\omega t)$$

$$\frac{1}{n} \sum_{t=1}^{n} \sum_{j=0}^{\infty} a(j) e(t - t) \left\{ \cos((t - j)\omega) \cos(j\omega) - \sin((t - j)\omega) \sin(j\omega) \right\}$$

$$= \frac{1}{n} \sum_{j=0}^{\infty} a(j) \cos(j\omega) \sum_{t=1}^{n} e(t - j) \cos((t - j)\omega)$$

$$- \frac{1}{n} \sum_{j=0}^{\infty} a(j) \sin(j\omega) \sum_{t=1}^{n} e(t - j) \sin((t - j)\omega). \tag{4.39}$$

Therefore,

$$\sup_{\omega} \frac{1}{n} \left| \sum_{t=1}^{n} X(t) \cos(\omega t) \right| \leq \sup_{\theta} \frac{1}{n} \left| \sum_{j=0}^{\infty} a(j) \cos(j\omega) \sum_{t=1}^{n} e(t-j) \cos((t-j)\omega) \right|$$

$$+ \sup_{\omega} \frac{1}{n} \left| \sum_{j=0}^{\infty} a(j) \sin(j\omega) \sum_{t=1}^{n} e(t-j) \sin((t-j)\omega) \right| \quad \text{a.s.}$$

$$\leq \frac{1}{n} \sum_{j=0}^{\infty} |a(j)| \sup_{\omega} \left| \sum_{t=1}^{n} e(t-j) \cos((t-j)\omega) \right|$$

$$+ \frac{1}{n} \sum_{j=0}^{\infty} |a(j)| \sup_{\omega} \left| \sum_{t=1}^{n} e(t-j) \sin((t-j)\omega) \right|.$$

Now taking expectation

$$E \left(\sup_{\omega} \frac{1}{n} \left| \sum_{t=1}^{n} X(t) \cos(\omega t) \right| \right)$$

$$\leq \frac{1}{n} \sum_{j=0}^{\infty} |a(j)| E \left(\sup_{\omega} \left| \sum_{t=1}^{n} e(t-j) \cos((t-j)\omega) \right| \right)$$

$$\leq \frac{1}{n} \sum_{j=0}^{\infty} |a(j)| \left\{ E \sup_{\theta} \left| \sum_{t=1}^{n} e(t-j) \cos((t-j)\omega) \right|^2 \right\}^{1/2}$$

$$+ \frac{1}{n} \sum_{j=0}^{\infty} |a(j)| \left\{ E \sup_{\theta} \left| \sum_{t=1}^{n} e(t-j) \sin((t-j)\omega) \right|^2 \right\}^{1/2}. \qquad (4.40)$$

The first term of (4.40)

$$\frac{1}{n} \sum_{j=0}^{\infty} |a(j)| \left\{ E \sup_{\theta} \left| \sum_{t=1}^{n} e(t-j) \cos((t-j)\omega) \right|^2 \right\}^{1/2}$$

$$\leq \frac{1}{n} \sum_{j=0}^{\infty} |a(j)| \left\{ n + \sum_{t=-(n-1)}^{n-1} E \left(\left| \sum_{m} e(m)e(m+t) \right| \right)^{1/2} \right\} \qquad (4.41)$$

where the sum $\sum_{t=-(n-1)}^{n-1}$ omits the term $t = 0$ and $\sum_m$ is over all such m such that $1 \leq m+t \leq n$, that is, $n - |t|$ terms (dependent on j). Similarly the second term of (4.41) can be bounded by the same. Since

$$E \left(\left| \sum_{m} e(m)e(m+t) \right| \right) \leq E \left(\left| \sum_{m} e(m)e(m+t) \right|^2 \right)^{1/2} = O(n^{1/2}), \qquad (4.42)$$

uniformly in j, the right-hand side of (4.41) is $O\left\{(n + n^{3/2})^{1/2}/n\right\} = O(n^{-1/4})$. Therefore, (4.40) is also $O(n^{-1/4})$. Let $M = n^3$, then $E\left(\sup_\omega \left|\sum_{t=1}^n X(t)\right.\right.$ $\left.\left.\cos(\omega t)\right|/n\right) \le O(n^{-3/2})$. Therefore using Borel Cantelli Lemma, it follows that

$$\sup_\omega \frac{1}{n}\left|\sum_{t=1}^n X(t)\cos(\omega t)\right| \to 0, \quad \text{a.s.}$$

Now for J, $n^3 < J \le (n+1)^3$,

$$\sup_\omega \sup_{n^3 < J < (n+1)^3} \left|\frac{1}{n^3}\sum_{t=1}^{n^3} X(t)\cos(\omega t) - \frac{1}{J}\sum_{t=1}^{J} X(t)\cos(\omega t)\right|$$

$$= \sup_\omega \sup_{n^3 < J < (n+1)^3} \left|\frac{1}{n^3}\sum_{t=1}^{n^3} X(t)\cos(\omega t) - \frac{1}{n^3}\sum_{t=1}^{J} X(t)\cos(\omega t)\right.$$

$$\left. + \frac{1}{n^3}\sum_{t=1}^{J} X(t)\cos(\omega t) - \frac{1}{J}\sum_{t=1}^{J} X(t)\cos(\omega t)\right|$$

$$\le \frac{1}{n^3}\sum_{t=n^3+1}^{(n+1)^3} |X(t)| + \sum_{t=1}^{(n+1)^3} |X(t)|\left(\frac{1}{n^3} - \frac{1}{(n+1)^3}\right) \quad \text{a.s.} \qquad (4.43)$$

The mean squared error of the first term is of the order $O\left((1/n^6) \times ((n+1)^3 - n^3)^2\right) = O(n^{-2})$ and the mean squared error of the second term is of the order $O\left(n^6 \times \left(((n+1)^3 - n^3)/n^6\right)^2\right) = O(n^{-2})$. Therefore, both terms converge to zero almost surely. That proves the lemma. $\qquad\qquad\qquad\qquad\qquad\qquad\qquad\qquad\qquad\qquad\qquad\square$

Proof of Lemma 4.4

Let $I'(\omega)$ and $I''(\omega)$ be the first and second derivatives of $I(\omega)$ with respect to ω. Expanding $I'(\widetilde{\omega})$ around ω^0 using Taylor series expansion.

$$I'(\widetilde{\omega}) - I'(\omega) = (\widetilde{\omega} - \omega^0)I''(\bar{\omega}), \qquad (4.44)$$

where $\bar{\omega}$ is a point on the line joining $\widetilde{\omega}$ and ω^0. Since $I'(\widetilde{\omega}) = 0$, (4.44) can be written as

$$n(\widetilde{\omega} - \omega^0) = \left[\frac{1}{n^2}I'(\omega^0)\right]\left[\frac{1}{n^3}I''(\bar{\omega})\right]^{-1}$$

It can be shown that $\lim_{n\to\infty} \frac{1}{n^2}I'(\omega^0) = 0$ a.s. and since $I''(\omega)$ is a continuous function of ω and $\widetilde{\omega} \to \omega^0$ a.s.

$$\lim_{n\to\infty} \frac{1}{n^3}I''(\bar{\omega}) = \frac{1}{24}(A^{0^2} + B^{0^2}) \neq 0. \tag{4.45}$$

Therefore, we have $n(\widetilde{\omega} - \omega^0) \to 0$ a.s. $\qquad\qquad\qquad \Box$

Appendix B

We here calculate the variance covariance matrix of $Q'(\boldsymbol{\theta}^0)\mathbf{D}$, present in (4.9). In this case, the error random variables $X(t)$ can be written as $\sum_{j=0}^{\infty} a(j)e(t-j)$. Note that

$$Q'(\boldsymbol{\theta}^0)\mathbf{D} = \left(\frac{1}{n^{\frac{1}{2}}}\frac{\partial Q(\boldsymbol{\theta})}{\partial A}, \frac{1}{n^{\frac{1}{2}}}\frac{\partial Q(\boldsymbol{\theta})}{\partial B}, \frac{1}{n^{\frac{3}{2}}}\frac{\partial Q(\boldsymbol{\theta})}{\partial \omega}\right)$$

and $\lim_{n\to\infty} \text{Var}(Q'(\boldsymbol{\theta}^0)\mathbf{D})) = \boldsymbol{\Sigma}$. In the following, we calculate Σ_{11} and Σ_{13}, where $\boldsymbol{\Sigma} = ((\Sigma_{ij}))$. Rest of the elements can be calculated similarly.

$$\Sigma_{11} = \lim_{n\to\infty} \text{Var}\left(\frac{1}{n^{\frac{1}{2}}}\frac{\partial Q(\boldsymbol{\theta})}{\partial A}\right) = \frac{1}{n}E\left[-2\sum_{t=1}^{n}X(t)\cos(\omega^0 t)\right]^2$$

$$= \lim_{n\to\infty} \frac{4}{n}E\left[\sum_{t=1}^{n}\sum_{j=0}^{\infty}a(j)e(t-j)\cos(\omega^0 t)\right]^3$$

$$= \lim_{n\to\infty} \frac{4}{n}E\left[\sum_{t=1}^{n}\sum_{j=0}^{\infty}a(j)e(t-j)\big(\cos(\omega^0(t-j))\cos(\omega^0 j)\right.$$

$$\left. - \sin(\omega^0(t-j))\sin(\omega^0 j)\big)\right]^2$$

$$= \lim_{n\to\infty} \frac{4}{n}E\left[\sum_{j=0}^{\infty}a(j)\cos(\omega^0 j)\sum_{t=1}^{n}e(t-j)\cos(\omega^0(t-j))\right.$$

$$\left. - \sum_{j=0}^{\infty}a(j)\sin(\omega^0 j)\sum_{t=1}^{n}e(t-j)\sin(\omega^0(t-j))\right]^2$$

$$= 4\sigma^2\left[\frac{1}{2}\left\{\sum_{j=0}^{\infty}a(j)\cos(\omega^0 j)\right\}^2 + \frac{1}{2}\left\{\sum_{j=0}^{\infty}a(j)\sin(\omega^0 j)\right\}^2\right]$$

$$= 2\sigma^2 \left| \sum_{j=0}^{\infty} a(j) e^{-ij\omega^0} \right|^2 = 2c(\omega^0).$$

$$\Sigma_{13} = \lim_{n \to \infty} \mathrm{Cov}\left(\frac{1}{n^{\frac{1}{2}}} \frac{\partial Q(\boldsymbol{\theta})}{\partial A}, \frac{1}{n^{\frac{3}{2}}} \frac{\partial Q(\boldsymbol{\theta})}{\partial \omega} \right)$$

$$= \lim_{n \to \infty} \frac{1}{n^2} E\left[-2 \sum_{t=1}^{n} X(t) \cos(\omega^0 t) \right] \left[2 \sum_{t=1}^{n} t X(t) \left(A^0 \sin(\omega^0 t) - B^0 \cos(\omega^0 t) \right) \right]$$

$$= \lim_{n \to \infty} -\frac{4}{n^2} E\left[\sum_{t=1}^{n} \sum_{j=0}^{\infty} a(j) e(t-j) \cos(\omega^0 t) \right]$$

$$\times \left[\sum_{t=1}^{n} t \sum_{j=0}^{\infty} a(j) e(t-j) \left\{ A^0 \sin(\omega^0 t) - B^0 \cos(\omega^0 t) \right\} \right]$$

$$= \lim_{n \to \infty} -\frac{4}{n^2} E\left[\sum_{t=1}^{n} \sum_{j=0}^{\infty} a(j) e(t-j) \left(\cos(\omega^0(t-j)) \cos(\omega^0 j) \right. \right.$$

$$\left. - \sin(\omega^0(t-j)) \sin(\omega^0 j) \right) \Big]$$

$$\times \left[\sum_{t=1}^{n} \sum_{j=0}^{\infty} t \, a(j) e(t-j) \left(A^0 \sin(\omega^0(t-j)) \cos(\omega^0 j) + A^0 \cos(\omega^0(t-j)) \sin(\omega^0 j) \right. \right.$$

$$\left. - B^0 \cos(\omega^0(t-j)) \cos(\omega^0 j) + B^0 \sin(\omega^0(t-j)) \sin(\omega^0 j) \right) \Big]$$

$$= \lim_{n \to \infty} -\frac{4}{n^2} E\left[\left(\sum_{j=0}^{\infty} a(j) \cos(\omega^0 j) \sum_{t=1}^{n} e(t-j) \cos(\omega^0(t-j)) \right. \right.$$

$$\left. - \sum_{j=0}^{\infty} a(j) \sin(\omega^0 j) \sum_{t=1}^{n} e(t-j) \sin(\omega^0(t-j)) \right)$$

$$\times \left(A^0 \sum_{j=0}^{\infty} a(j) \cos(\omega^0 j) \sum_{t=1}^{n} t \, e(t-j) \sin(\omega^0(t-j)) \right.$$

$$+ A^0 \sum_{j=0}^{\infty} a(j) \sin(\omega^0 j) \sum_{t=1}^{n} t \, e(t-j) \cos(\omega^0(t-j))$$

$$- B^0 \sum_{j=0}^{\infty} a(j) \cos(\omega^0 j) \sum_{t=1}^{n} t \, e(t-j) \cos(\omega^0(t-j))$$

$$+ B^0 \sum_{j=0}^{\infty} a(j) \sin(\omega^0 j) \sum_{t=1}^{n} t \, e(t-j) \sin(\omega^0(t-j)) \right) \Big]$$

$$= -4\left[A^0\frac{1}{4}\left\{\sum_{j=0}^{\infty}a(j)\cos(\omega^0 j)\right\}\right.$$

$$\times\left\{\sum_{j=0}^{\infty}a(j)\sin(\omega^0 j)\right\}\sigma^2 - B^0\frac{1}{4}\left\{\sum_{j=0}^{\infty}a(j)\cos(\omega^0 j)\right\}^2$$

$$\left. - A^0\left\{\sum_{j=0}^{\infty}a(j)\cos(\omega^0 j)\right\}\left\{\sum_{j=0}^{\infty}a(j)\sin(\omega^0 j)\right\} - B^0\frac{1}{4}\left\{\sum_{j=0}^{\infty}a(j)\sin(\omega^0 j)\right\}^2\right]$$

$$= \sigma^2 B^0\left|\sum_{j=0}^{\infty}a(j)e^{ij\omega^0}\right|^2 = B^0\sigma^2 c(\omega^0). \qquad \square$$

References

1. Kundu, D. (1993). Asymptotic theory of least squares estimators of a particular non-linear regression model. *Statistics and Probability Letters, 18*, 13–17.
2. Jennrich, R. I. (1969). Asymptotic properties of the non-linear least squares estimators. *Annals of Mathematical Statistics, 40*, 633–643.
3. Wu, C. F. J. (1981). Asymptotic theory of non-linear least squares estimation. *Annals Statistics, 9*, 501–513.
4. Whittle, P. (1952). The simultaneous estimation of a time series Harmonic component and covariance structure. *Trabajos de Estadistica, 3*, 43–57.
5. Hannan, E. J. (1971). Non-linear time series regression. *Journal of Applied Probability, 8*, 767–780.
6. Hannan, E. J. (1973). The estimation of frequency. *Journal of Applied Probability,10*, 510–519.
7. Walker, A. M. (1971). On the estimation of a harmonic component in a time series with stationary independent residuals. *Biometrika, 58*, 21–36.
8. Rice, J. A., & Rosenblatt, M. (1988). On frequency estimation. *Biometrika, 75*, 477–484.
9. Quinn, B. G., & Fernandes, J. M. (1991). A fast efficient technique for the estimation of frequency. *Biometrika,78*, 489–497.
10. Kundu, D. (1997). Estimating the number of sinusoids in additive white noise. *Signal Processing, 56*, 103–110.
11. Quinn, B. G. (1994). Estimating frequency by interpolation using Fourier coefficients. *IEEE Transactions on Signal Processing, 42*, 1264–1268.
12. Kundu, D., & Mitra, A. (1996). Asymptotic theory of least squares estimates of a non-linear time series regression model. *Communication in Statistics Theory and Methods, 25*, 133–141.
13. Irizarry, R. A. (2002). Weighted estimation of harmonic components in a musical sound signal. *Journal of Time Series Analysis, 23*, 29–48.
14. Nandi, S., Iyer, S. K., & Kundu, D. (2002). Estimating the frequencies in presence of heavy tail errors. *Statistics and Probability Letters, 58*, 265–282.
15. Prasad, A., Kundu, D., & Mitra, A. (2008). Sequential estimation of the sum of sinusoidal model parameters. *Journal of Statistical Planning and Inference, 138*, 1297–1313.
16. Kundu, D., Bai, Z. D., Nandi, S., & Bai, L. (2011). Super efficient frequency Estimation. *Journal of Statistical Planning and Inference, 141*(8), 2576–2588.
17. Fuller, W. A. (1976). *Introduction to statistical time series*. New York: John Wiley and Sons.
18. Samorodnitsky, G., & Taqqu, M. S. (1994). *Stable non-Gaussian random processes; stochastic models with infinite variance*. New York: Chapman and Hall.

19. Bai, Z. D., Rao, C. R., Chow, M., & Kundu, D. (2003). An efficient algorithm for estimating the parameters of superimposed exponential signals. *Journal of Statistical Planning and Inference, 110*, 23–34.
20. Nandi, S., & Kundu, D. (2006). A fast and efficient algorithm for estimating the parameters of sum of sinusoidal model. *Sankhya, 68*, 283–306.
21. Bai, Z. D., Chen, X. R., Krishnaiah, P. R., & Zhao, L. C. (1987). Asymptotic properties of EVLP estimators for superimposed exponential signals in noise. Technical Report. 87–19, CMA, U. Pittsburgh.
22. Kundu, D., & Mitra, A. (1997). Consistent methods of estimating sinusoidal frequencies; a non iterative approach. *Journal of Statistical Computation and Simulation, 58*, 171–194.
23. Kundu, D. (1994). A modified Prony algorithm for sum of damped or undamped exponential signals. *Sankhya, 56*, 524–544.
24. Kannan, N., & Kundu, D. (1994). On modified EVLP and ML methods for estimating superimposed exponential signals. *Signal Processing, 39*, 223–233.
25. Truong-Van, B. (1990). A new approach to frequency analysis with amplified harmonics. *Journal of the Royal Statistical Society Series B, 52*, 203–221.

Chapter 5
Estimating the Number of Components

5.1 Introduction

In the previous two chapters, we have discussed different estimation procedures of model (3.1) and properties of these estimators. In all these developments, it has been assumed that the number of components 'p' is known in advance. But in practice estimation of p is also a very important problem. Although, during the last 35 to 40 years extensive work has been done in estimating the frequencies of model (3.1), not that much attention has been paid in estimating the number of components p.

The estimation of 'p' can be considered as a model selection problem. Consider the class of models

$$\mathscr{M}_k = \left\{ \mu_k; \mu_k(t) = \sum_{j=1}^{k} A_j \cos\left(\omega_j t\right) + B_j \sin\left(\omega_j t\right) \right\}; \quad \text{for} \quad k = 1, 2, \ldots.$$

(5.1)

Based on the data $\{y(t); t = 1, 2 \cdots, n\}$, estimating '$p$' is equivalent to find $\widehat{p}$, so that $\mathscr{M}_{\widehat{p}}$ becomes the 'best' fitted model to the data. Therefore, any model selection method can be used in principle to choose p.

The most intuitive and natural estimator of p is the number of peaks of the periodogram function of the data as defined in (1.5). Consider the following examples.

Example 5.1 The data $\{y(t), t = 1, \ldots, n\}$ are obtained from model (3.1) with model parameters;

$$p = 2, \quad A_1 = A_2 = 1.0, \quad \omega_1 = 1.5, \quad \omega_2 = 2.0.$$

(5.2)

The error random variables $X(1), \ldots, X(n)$ are i.i.d. normal random variables with mean 0 and variance 1. The periodogram function is plotted in Fig. 5.1, and it is immediate from the plot that the number of components is 2.

D. Kundu and S. Nandi, *Statistical Signal Processing*, SpringerBriefs in Statistics, DOI: 10.1007/978-81-322-0628-6_5, © The Author(s) 2012

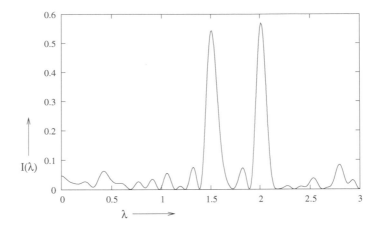

Fig. 5.1 The periodogram plot of the data obtained from model (5.2)

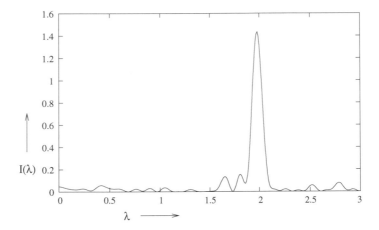

Fig. 5.2 The periodogram plot of the data obtained from model (5.3)

Example 5.2 The data $\{y(t), t = 1, \ldots, n\}$ are obtained from model (3.1) with model parameters;

$$p = 2, \quad A_1 = A_2 = 1.0, \quad \omega_1 = 1.95, \quad \omega_2 = 2.0. \tag{5.3}$$

The error random variables are same as in Example 5.1. The periodogram is plotted in Fig. 5.2. It is not clear from Fig. 5.2 that $p = 2$.

Example 5.3 The data $\{y(t); t = 1, \ldots, n\}$ are obtained from model (3.1) with the same model parameters as in Example 5.1, but the errors are i.i.d. normal random

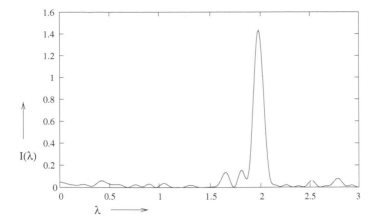

Fig. 5.3 The periodogram plot of the data obtained from model (5.2) with error variance 5.

variables with mean zero and variance 5. The periodogram function is presented in Fig. 5.3. It is not clear again from the periodogram plot that $p = 2$.

The above examples reveal that when frequencies are very close to each other or if the error variance is high, it may not be possible to detect the number of components from the periodogram plot of the data. Different methods have been proposed to detect the number of components of model (3.1). All the methods can be broadly classified into three different categories namely; (a) likelihood ratio approach, (b) cross validation method, and (c) information theoretic criterion. In this chapter, we provide a brief review of different methods. Throughout this chapter without loss of generality, one can assume that

$$A_1^2 + B_1^2 > A_2^2 + B_2^2 > \cdots > A_p^2 + B_p^2.$$

5.2 Likelihood Ratio Approach

In estimating p of model (3.1), one of the natural procedures is to use a test of significance for each additional term as it is introduced in the model. Fisher [1] considered this as a simple testing of hypothesis problem. Such a test can be based on the well-known 'maximum likelihood ratio', that is, the ratio of the maximized likelihood for k terms of model (3.1) to the maximized likelihood for $(k-1)$ terms of model (3.1). If this quantity is large, it provides evidence that the kth term is needed in the model.

The problem can be formulated as follows:

$$H_0 : p = p_0 \quad \text{against} \quad H_1 : p = p_1, \tag{5.4}$$

where $p_1 > p_0$. Based on the assumption that the error random variables follow i.i.d. normal distribution with mean 0 and variance σ^2, the maximized log-likelihood for fixed σ^2 can be written as

$$\text{constant} - \frac{n}{2} \ln \sigma^2 - \frac{1}{2\sigma^2} \sum_{t=1}^{n} \left[y(t) - \sum_{k=1}^{p_0} \{ \widehat{A}_{k,p_0} \cos \left(\widehat{\omega}_{k,p_0} t \right) \right.$$

$$\left. + \widehat{B}_{k,p_0} \sin \left(\widehat{\omega}_{k,p_0} t \right) \} \right]^2,$$

here $\widehat{A}_{k,p_0}, \widehat{B}_{k,p_0}, \widehat{\omega}_{k,p_0}$ are the MLEs of $A_{k,p_0}, B_{k,p_0}, \omega_{k,p_0}$, respectively, based on the assumption that $p = p_0$. The unconstrained maximized log-likelihood is then

$$l_{p_0} = \text{constant} - \frac{n}{2} \ln \widehat{\sigma}_{p_0}^2 - \frac{n}{2}, \tag{5.5}$$

here

$$\widehat{\sigma}_{p_0}^2 = \frac{1}{n} \sum_{t=1}^{n} \left[y(t) - \sum_{k=1}^{p_0} \{ \widehat{A}_{k,p_0} \cos \left(\widehat{\omega}_{k,p_0} t \right) + \widehat{B}_{k,p_0} \sin \left(\widehat{\omega}_{k,p_0} t \right) \} \right]^2.$$

Therefore, the likelihood ratio test takes the following form: rejects H_0 if L is large, where

$$L = \frac{\widehat{\sigma}_{p_0}^2}{\widehat{\sigma}_{p_1}^2} = \frac{\sum_{t=1}^{n} \left[y(t) - \sum_{k=1}^{p_0} \{ \widehat{A}_{k,p_0} \cos \left(\widehat{\omega}_{k,p_0} t \right) + \widehat{B}_{k,p_0} \sin \left(\widehat{\omega}_{k,p_0} t \right) \} \right]^2}{\sum_{t=1}^{n} \left[y(t) - \sum_{k=1}^{p_1} \{ \widehat{A}_{k,p_1} \cos \left(\widehat{\omega}_{k,p_1} t \right) + \widehat{B}_{k,p_1} \sin \left(\widehat{\omega}_{k,p_1} t \right) \} \right]^2}.$$
$$\tag{5.6}$$

To find the critical point of the above test procedure, one needs to obtain the exact/ asymptotic distribution of L under the null hypothesis. It seems finding the exact/ asymptotic distribution of L is a difficult problem.

Quinn [2] obtained the distribution of L as defined in (5.6) under the assumptions: (a) errors are i.i.d. normal random variables, with mean 0 and variance σ^2 and (b) frequencies are of the form $2\pi j/n$, where $1 \leq j \leq (n-1)/2$. If the frequencies are of the form (b), Quinn [2] showed that in this case L is of form:

$$L = \frac{\sum_{t=1}^{n} y(t)^2 - J_{p_0}}{\sum_{t=1}^{n} y(t)^2 - J_{p_1}}, \tag{5.7}$$

where J_k is the sum of the k largest elements of $\{I(\omega_j); \omega_j = 2\pi j/n, 1 \leq j \leq (n-1)/2\}$, and $I(\omega)$ is the periodogram function of the data sequence $\{y(t); t = 1, \ldots, n\}$, as defined in (1.5). The likelihood ratio statistic L as defined in (5.7) can also be written as

$$L = \frac{\sum_{t=1}^{n} y(t)^2 - J_{p_0}}{\sum_{t=1}^{n} y(t)^2 - J_{p_1}} = \frac{1}{1 - G_{p_0,p_1}}, \tag{5.8}$$

where

$$G_{p_0,p_1} = \frac{J_{p_1} - J_{p_0}}{\sum_{t=1}^{n} y(t)^2 - J_{p_0}}. \tag{5.9}$$

When $p_0 = 0$ and $p_1 = 1$,

$$G_{0,1} = \frac{J_1}{\sum_{t=1}^{n} y(t)^2}, \tag{5.10}$$

and it is the well-known Fisher's g-statistic.

Finding the distribution of L is equivalent to finding the distribution of G_{p_0,p_1}. Quinn and Hannan [3] provided the approximate distribution of G_{p_0,p_1}, which is quite complicated in nature and may not have much practical importance. The problem becomes more complicated when the frequencies are not in the form of (b) as defined above. Some attempts have been made to simplify the distribution of G_{p_0,p_1} by Quinn and Hannan [3] under the assumption of i.i.d. normal error. It is not further pursued here.

5.3 Cross Validation Method

Cross validation method is a model selection technique and it can be used in a fairly general setup. The basic assumption of cross validation technique is that there exists an M, such that $1 \leq k \leq M$ for the models defined in (5.1). The cross validation method can be described as follows: For a given k, such that $1 \leq k \leq M$, remove jth observation from $\{y(1), \ldots, y(n)\}$, and estimate $y(j)$, say $\widehat{y}_k(j)$, based on the model assumption $\mathcal{M}_k$ and $\{y(1), \ldots, y(j-1), y(j+1), \ldots, y(n)\}$. Compute the cross validatory error for the kth model as

$$CV(k) = \sum_{t=1}^{n} (y(t) - \widehat{y}_k(t))^2; \quad \text{for}\ \ k = 1, \ldots, M. \tag{5.11}$$

Choose $\widehat{p}$ as the estimate of p, if

$$CV(\widehat{p}) < \{CV(1), \ldots, CV((\widehat{p}-1), CV((\widehat{p}+1), \ldots, CV(M)\}.$$

Cross validation method has been used quite extensively in model selection as it is well known that the small sample performance of cross validation technique is very good, although it usually does not produce consistent estimator of the model order.

Rao [4] proposed to use the cross validation technique to estimate the number of components in a sinusoidal model. The author did not provide any explicit method to compute $\widehat{y}_k(j)$, based on the observations $\{y(1), \ldots, y(j-1), y(j+1), \ldots, y(n)\}$ and most of the estimation methods are based on the fact that the data are equispaced.

Kundu and Kundu [5] first provided modified EVLP method and some of its generalizations to estimate consistently the amplitudes and frequencies of a sinusoidal signal based on the assumption that the errors are i.i.d. random variables with mean zero and finite variance. The modified EVLP method has been further modified by Kundu and Mitra [6] by using both the forward and backward data, and it has been used quite effectively to estimate p of model (3.1) by using cross validation technique. Extensive simulation results suggest that the cross validation technique works quite well for small sample sizes and for large error variances, although for large sample sizes the performance is not that satisfactory. For large sample sizes, the cross validation technique is computationally very demanding, hence it is not recommended.

In both the likelihood ratio approach and the cross validation approach, it is important that the errors are i.i.d. random variables with mean zero and finite variance. It is not immediate how these methods can be modified for stationary errors.

5.4 Information Theoretic Criteria

Different information theoretic criteria such as AIC of Akaike [7, 8], BIC of Schwartz [9] or Rissanen [10], EDC of Bai et al. [11] have been used quite successfully in different model selection problems. AIC, BIC, EDC, and their several modifications have been used to detect the number of components of model (3.1). All the information theoretic criteria are based on the following assumption that the maximum model order can be M, and they can be put in the general framework as follows: For the kth order model, define

$$ITC(k) = f\left(\widehat{\sigma}_k^2\right) + N(k)\, c(n); \quad 1 \le k \le M, \tag{5.12}$$

here $\widehat{\sigma}_k^2$ is the estimated error variance and $N(k)$ denotes the number of parameters, both based on the assumption that the model order is k, $f(\cdot)$ is an increasing and $c(\cdot)$ is a monotone function of n. The quantity $N(k)c(n)$ is known as the penalty, and for fixed n, it increases with k. Depending on different information theoretic criteria, $f(\cdot)$ and $c(\cdot)$ change. Choose $\widehat{p}$ as an estimate of p, if

$$ITC\left(\widehat{p}\right) < \{ITC(1), \ldots, ITC\left((\widehat{p}-1)\right), ITC\left((\widehat{p}+1)\right), \ldots, ITC(M)\}.$$

The main focus of the different information theoretic criteria is to choose properly the functions $f(\cdot)$ and $c(\cdot)$.

5.4.1 Rao's Method

Rao [4] proposed different information theoretic criteria to detect the number of sinusoidal components based on the assumption that the errors are i.i.d. mean zero normal random variables. Based on the above error assumption, AIC takes the following form

$$AIC(k) = n \ln R_k + 2\,(7k),\qquad(5.13)$$

here R_k denotes the minimum value of

$$\sum_{t=1}^{n}\left(y(t) - \sum_{j=1}^{k}\left(A_j\cos\left(\omega_j t\right) + B_j\sin\left(\omega_j t\right)\right)\right)^2,\qquad(5.14)$$

and the minimization of (5.14) is performed with respect to $A_1, \ldots, A_k, B_1, \ldots, B_k,$ $\omega_1, \ldots, \omega_k$. '$7k$' denotes the number of parameters when the number of components is k.

Under the same assumption, BIC takes the form

$$BIC(k) = n \ln R_k + (7k)\frac{1}{2}\ln n,\qquad(5.15)$$

and EDC takes the form;

$$EDC(k) = n \ln R_k + (7k)\,c(n),\qquad(5.16)$$

here $c(n)$ satisfies the following conditions;

$$\lim_{n\to\infty}\frac{c(n)}{n} = 0 \quad\text{and}\quad \lim_{n\to\infty}\frac{c(n)}{\ln\ln n} = \infty.\qquad(5.17)$$

EDC is a very flexible criterion, and BIC is a special case of EDC. Several $c(n)$ satisfy (5.17). For example, $c(n) = n^a$, for $a < 1$ and $c(n) = (\ln\ln n)^b$, for $b > 1$ satisfy (5.17).

Although Rao [4] proposed to use information theoretic criteria to detect the number of components of model (3.1), he did not provide any practical implementation procedure particularly, the computation of R_k. He suggested to compute R_k by minimizing (5.14) with respect to the unknown parameters, which may not be very simple, as it has been observed in Chap. 3.

Kundu [12] suggested a practical implementation procedure of the method proposed by Rao [4], and performed extensive simulation studies to compare different methods for different models, for different error variances, and for different choices of $c(n)$. It is further observed that AIC does not provide consistent estimate of the model order. Although, for small sample sizes the performance of AIC is good, for large sample sizes it has a tendency to overestimate the model order. Among the different choices of $c(n)$ for EDC criterion, it is observed that BIC performs quite well.

5.4.2 Sakai's Method

Sakai [13] considered the problem of estimating p of model (3.1) under the assumptions that the errors are i.i.d. normal random variables with mean zero and the frequencies can be Fourier frequencies only. He has re-formulated the problem as follows. Consider the model

$$y(t) = \sum_{j=0}^{M} v_j \left(A_j \cos \left(\omega_j t \right) + B_j \left(\sin(\omega_j t) \right) \right) + X(t); \quad t = 1, \ldots, n, \quad (5.18)$$

here $\omega_j = 2\pi j/n$, and $M = n/2$ or $M = (n-1)/2$ depending on whether n is even or odd. The indicator function v_j is such that

$$v_j = \begin{cases} 1 & \text{if } j \text{ th component is present} \\ 0 & \text{if } j \text{ th component is absent.} \end{cases}$$

Sakai [13] proposed the following information theoretic like criterion as follows

$$SIC(v_1, \ldots, v_M) = \ln \widehat{\sigma}^2 + \frac{2(\ln n + \gamma - \ln 2)}{n}(v_1 + \cdots + v_M), \quad (5.19)$$

here $\gamma (\approx = 0.577)$ is the Euler's constant and

$$\widehat{\sigma}^2 = \frac{1}{n} \sum_{t=1}^{n} y(t)^2 - \frac{4}{n} \sum_{k=1}^{M} I(\omega_k) v_k,$$

here $I(\cdot)$ is the periodogram function of $\{y(t); t = 1, \ldots, n\}$. Now for all 2^M possible choices of $(v_1, \ldots, v_M)$, choose that combination for which $SIC(v_1, \ldots, v_M)$ is minimum.

5.4.3 Quinn's Method

Quinn [14] considered the same problem under the assumptions that the sequence of error random variables $\{X(t)\}$ is stationary and ergodic with mean zero and finite variance. It is further assumed that the frequencies are of the form $2\pi j/n$, for $1 \le j \le (n-1)/2$. Under the above assumptions, Quinn [14] proposed an information theoretic like criterion as follows. Let

$$QIC(k) = n \ln \widehat{\sigma}_k^2 + 2k\, c(n) \tag{5.20}$$

where

$$\widehat{\sigma}_k^2 = \frac{1}{n}\left(\sum_{t=1}^{n} y(t)^2 - J_k\right),$$

and J_k is same as defined in (5.7). The penalty function $c(n)$ is such that it satisfies

$$\lim_{n\to\infty} \frac{c(n)}{n} = 0. \tag{5.21}$$

Then the number of sinusoids p is estimated as the smallest value of $k \geq 0$, for which $QIC(k) < QIC(k+1)$. Using the results of An et al. [15], it has been shown that if $\widehat{p}$ is an estimate of p, then $\widehat{p}$ converges to p almost surely. Although this method provides a consistent estimate of the number of sinusoidal components, it is not known how the method behaves for small sample sizes. Simulation experiments need to be done to verify the performance of this method.

5.4.4 Wang's Method

Wang [16] also considered this problem of estimating p under a more general condition than Rao [4] or Quinn [14]. Wang [16] assumed the same error assumptions as those of Quinn [14], but the frequencies need not be restricted to only Fourier frequencies. The method of Wang [16] is very similar to the method proposed by Rao [4], but the main difference is in the estimation procedure of the unknown parameters for the kth order model. The information criterion of Wang [16] can be described as follows. For the kth order model consider

$$WIC(k) = n \ln \widehat{\sigma}_k^2 + k\, c(n), \tag{5.22}$$

here $\widehat{\sigma}_k^2$ is the estimated error variance and $c(n)$ satisfies the same condition as (5.21). Although Rao [4] did not provide any efficient estimation procedure of the unknown parameters for the kth order model, Wang [16] suggested to use the following estimation procedure of the unknown frequencies. Let $\Omega_1 = (-\pi, \pi]$, and $\widehat{\omega}_1$ is the argument maximizer of the periodogram function (1.5) over Ω_1. For $j > 1$, if Ω_{j-1} and $\widehat{\omega}_{j-1}$ are defined, then

$$\Omega_j = \Omega_{j-1} \setminus (\widehat{\omega}_{j-1} - u_n, \widehat{\omega}_{j-1} + u_n),$$

and $\widehat{\omega}_j$ is obtained as the argument maximizer of the periodogram function over Ω_j, where $u_n > 0$ and satisfies the conditions

$$\lim_{n \to \infty} u_n = 0, \quad \text{and} \quad \lim_{n \to \infty} (n \ln n)^{1/2} u_n = \infty.$$

Estimate p as the smallest value of $k \geq 0$, such that $WIC(k) < WIC(k+1)$. If

$$\liminf_{n \to \infty} \frac{c(n)}{\ln n} > C > 0,$$

then an estimator of p obtained by this method is a strongly consistent estimator of p, see Wang [16]. Although Wang's method is known to provide a consistent estimate of p, it is not known how this method performs for small sample sizes. Moreover, he did not mention about the practical implementation procedure of his method, mainly how to choose u_n or $c(n)$ for a specific case. It is very clear that the performance of this procedure heavily depends on them.

Kavalieries and Hannan [17] discussed some practical implementation procedure of Wang's method. They have suggested a slightly different estimation procedure of the unknown parameters than that of Wang [16]. It is not pursued further here, interested readers may have a look at that paper.

5.4.5 Kundu's Method

Kundu [18] suggested the following simple estimation procedure of p. If M denotes the maximum possible model order, then for some fixed $L > 2M$, consider data matrix $\mathbf{A}_L$:

$$\mathbf{A}_L = \begin{pmatrix} y(1) & \cdots & y(L) \\ \vdots & \ddots & \vdots \\ y(n-L+1) & \cdots & y(n) \end{pmatrix}. \tag{5.23}$$

Let $\widehat{\sigma}_1^2 > \cdots > \widehat{\sigma}_L^2$ be the L eigenvalues of the $L \times L$ matrix $\mathbf{R}_n = \mathbf{A}_L^T \mathbf{A}_L / n$. Consider

$$KIC(k) = \widehat{\sigma}_{2k+1}^2 + k\, c(n), \tag{5.24}$$

here $c(n) > 0$, satisfying the following two conditions;

$$\lim_{n \to \infty} c(n) = 0, \quad \text{and} \quad \lim_{n \to \infty} \frac{c(n)\sqrt{n}}{(\ln \ln n)^{1/2}} = \infty. \tag{5.25}$$

Now choose that value of k as an estimate of p for which $KIC(k)$ is minimum.

Under the assumption of i.i.d. errors, the author proved the strong consistency of the above procedure. Moreover, the probability of wrong detection has also been obtained in terms of linear combination of chi-square variables. Extensive simulations have been performed to check the effectiveness of the proposed method and to find proper $c(n)$. It is observed that $c(n) = 1/\sqrt{\ln n}$ works quite well, although no theoretical justification has been provided.

Kundu [19] in a subsequent paper discussed the choice of $c(n)$. He has considered a slightly different criterion than (5.24), and the new criterion takes the form

$$KIC(k) = \ln\left(\widehat{\sigma}_{2k+1}^2 + 1\right) + k \, c(n), \tag{5.26}$$

here $c(n)$ satisfies the same assumptions as in (5.25). In this case also, the bounds on the probability of overestimation and probability of underestimation are obtained theoretically, and extensive simulation results suggest that these theoretical bounds match very well with the simulated results.

The natural question is how to choose proper $c(n)$. It is observed, see Kundu [19], that the probability of over estimates and probability of under estimates depend on the eigenvalues of the dispersion matrix of the asymptotic distribution of Vec($\mathbf{R}_n$). Here Vec($\cdot$) of a $k \times k$ matrix is a $k^2 \times 1$ vector stacking the columns one below the other. The main idea to choose the proper penalty function from a class of penalty functions which satisfy (5.25) is to choose that penalty function for which the theoretical bound of probability of wrong selection is minimum. These theoretical bounds depend on the parameters, and without knowing the unknown parameters it is not possible to calculate these theoretical bounds. The natural way to estimate these bound is by replacing the true parameter values by their estimates. Kundu [19] suggested a bootstrap-like technique to estimate these probabilities based on the observed sample, and then uses it in choosing the proper penalty function from a class of penalty functions.

The second question naturally arises how to choose the class of penalty functions. The suggestion is as follows. Take any particular class of reasonable size maybe around 10 or 12 where all of them should satisfy (5.25) but they should converge to zero at varying rates (from very low to very high). Obtain the probability of wrong detection for all these penalty functions and compute the minimum of these values. If the minimum itself is high, it indicates that the class is not good, otherwise it is fine. Simulation results indicate that the method works very well for different sample sizes and for different error variances. The major drawback of this method is that it has been proposed when the errors are i.i.d. random variables, it is not immediate how the method can be modified for the correlated errors.

5.5 Conclusions

In this chapter, we have discussed different methods of estimating the number of components in a multiple sinusoidal model. This problem can be formulated as a model selection problem, hence any model selection procedure which is available in the literature can be used for this purpose. We have provided three different approaches, comparison of different methods is not available in the literature. We believe it is still an open problem, more work is needed along this direction.

References

1. Fisher, R. A. (1929). Tests of significance in harmonic Analysis. *Proceedings of the Royal Society London Series A, 125*, 54–59.
2. Quinn, B.G. (1986). Testing for the presence of sinusoidal components. *Journal of Applied Probability, 23*(A), 201–210.
3. Quinn, B. G., & Hannan, E. J. (2001). *The estimation and tracking of frequency*. New York: Cambridge University Press.
4. Rao, C. R. (1988). Some results in signal detection. In S. S. Gupta & J. O. Berger (Eds.), *Decision theory and related topics IV* (pp. 319–332). New York: Springer.
5. Kundu, D., & Kundu, R. (1995). Consistent estimates of super imposed exponential signals when observations are missing. *Journal of Statistical Planning and Inference, 44*, 205–218.
6. Kundu, D., & Mitra, A. (1995). Consistent method of estimating the superimposed exponential signals. *Scandinavian Journal of Statistics, 22*, 73–82.
7. Akaike, H. (1969). Fitting autoregressive models for prediction. *Annals of the Institute of Statistical Mathematics, 21*, 243–247.
8. Akaike, H. (1970). Statistical predictor identification. *Annals of the Institute of Statistical Mathematics, 22*, 203–217.
9. Schwartz, S. C. (1978). Estimating the dimension of a model. *Annals of Statistics, 6*, 461–464.
10. Rissanen, J. (1978). Modeling by shortest data description. *Automatica, 14*, 465–471.
11. Bai, Z. D., Krishnaiah, P. R., & Zhao, L. C. (1986). On the detection of the number of signals in the presence of white noise. *Journal of Multivariate Analysis, 20*, 1–25.
12. Kundu, D. (1992). Detecting the number of signals for undamped exponential models using information theoretic criteria. *Journal of Statistical Computation and Simulation, 44*, 117–131.
13. Sakai, H. (1990). An application of a BIC-type method to harmonic analysis and a new criterion for order determination of an error process. *IEEE Transaction of Acoustics Speech Signal Process, 38*, 999–1004.
14. Quinn, B. G. (1989). Estimating the number of terms in a sinusoidal regression. *Journal of Time Series Analysis, 10*, 71–75.
15. An, H.-Z., Chen, Z.-G., & Hannan, E. J. (1983). The maximum of the periodogram. *Journal of Multivariate Analysis, 13*, 383–400.
16. Wang, X. (1993). An AIC type estimator for the number of cosinusoids. *Journal of Time Series Analysis, 14*, 433–440.
17. Kavalieris, L., & Hannan, E. J. (1994). Determining the number of terms in a trigonometric regression. *Journal of Time Series Analysis, 15*, 613–625.
18. Kundu, D. (1997). Estimating the number of sinusoids in additive white noise. *Signal Processing, 56*, 103–110.
19. Kundu, D. (1998). Estimating the number of sinusoids and its performance analysis. *Journal of Statistical Computation and Simulation, 60*, 347–362.

Chapter 6
Real Data Example

6.1 Introduction

In this chapter, we analyze four data sets using multiple sinusoidal model. These data sets are namely (i) a segment of ECG signal of a healthy human being, (ii) the well-known variable star data, (iii) a short duration voiced speech signal, the "uuu" data, and (iv) the airline passenger data. Periodogram function defined in (1.5), as well as $R_1(\omega)$ defined in (3.39), can be used for initial identification of the number of components. Once the initial estimates are found, the unknown parameters are estimated using sequential estimation procedure. For the estimation of the number of components p, information theoretic criteria are used.

6.2 ECG Data

The data represents a segment of ECG signal of a healthy human being and is plotted in Fig. 1.1. The data set contains 512 points, and it has been analyzed by Prasad et al. [1]. The data are first mean corrected and scaled by the square root of the estimated variance of $\{y(t)\}$. The $S(\omega)$ function, as defined in (3.55), in the interval $(0, \pi)$ is plotted in Fig. 6.1. This gives an idea about the number of frequencies present in the data. We observe that the total number of frequencies is quite large. It is not easy to obtain an estimate of p from this figure. Apart from that all the frequencies may not be visible, depending on the magnitudes of some of the dominant frequencies and error variance. In fact, $\widehat{p}$ is much larger than what Fig. 6.1 reveals. The BIC has been used to estimate p. We estimate the unknown parameters of the multiple sinusoidal model sequentially for $k = 1, \ldots, 100$. For each k, the residual series is approximated by an AR process and the corresponding parameters are estimated. Here, k represents the number of components. Let ar_k be the number of AR parameters in the AR model fitted to the residuals when k sinusoidal components are estimated and $\widehat{\sigma}_k^2$ be the estimated error variance. Then minimize BIC(k) for this class of models for estimating p, which takes the following form in this case;

D. Kundu and S. Nandi, *Statistical Signal Processing*, SpringerBriefs in Statistics,
DOI: 10.1007/978-81-322-0628-6_6, © The Author(s) 2012

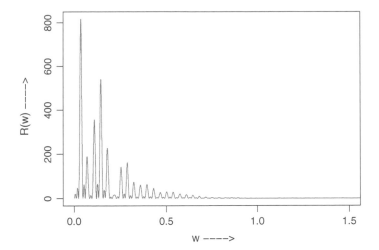

Fig. 6.1 The plot of the $S(\omega)$ function of the ECG signal data

$$\text{BIC}(k) = n \log \widehat{\sigma}_k^2 + \frac{1}{2}(3k + ar_k + 1) \log n.$$

The BIC(k) values are plotted in Fig. 6.2 for $k = 75, \ldots, 85$ and at $k = 78$, the BIC(k) takes its minimum value, therefore, we estimate p as $\widehat{p} = 78$. We estimate the unknown parameters using sequential method described in Sect. 3.12 as $\widehat{p}$ is quite large and simultaneous estimation might be a problem. With the estimated $\widehat{p}$, we plug-in the other estimates of the linear parameters and frequencies and obtain the fitted values $\{\widehat{y}(t); t = 1, \ldots, n\}$. They are plotted in Fig. 6.3 along with their observed values. The fitted values match reasonably well with the observed one. The residual sequence satisfies the assumption of stationarity.

6.3 Variable Star Data

The variable star data is an astronomical data and widely used in time series literature. This data set represents the daily brightness of a variable star on 600 successive midnights. The data is collected from Time Series Library of StatLib (http://www.stat. cmu.edu; Source: Rob J. Hyndman). The observed data is displayed in Fig. 1.2 and its periodogram function in Fig. 6.4. Initial inspection of the periodogram function gives an idea of the presence of two frequency components in the data, resulting in two sharp separate peaks in the periodogram plot. With $\widehat{p} = 2$, once we estimate the frequencies and the amplitudes and obtain the residual series, the periodogram plot of the resulting residual series gives evidence of the presence of another significant frequency. The not so dominating third component is not visible in the periodogram plot of the original series. This is due to the fact that the first two components are

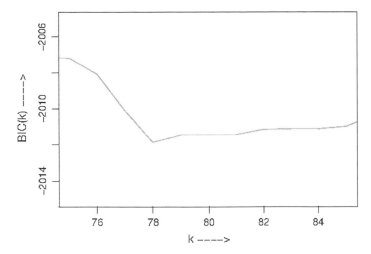

Fig. 6.2 The BIC values for different number of components

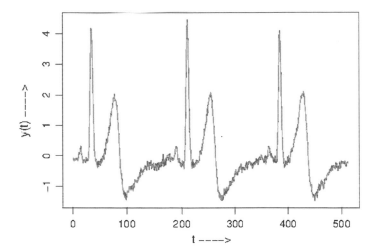

Fig. 6.3 The plot of the observed (*red*) and fitted values (*green*) of the ECG signal

dominant in terms of the large absolute values of their associated amplitudes. In addition, the first one is very close to the third one as compared to the available data points to distinguish them. Therefore, we take $\widehat{p} = 3$ and estimate the unknown parameters. The estimated point estimates are listed in Table 6.1. The observed (solid line) and the estimated values (dotted line) are plotted in Fig. 6.5, and it is not possible to distinguish them. So, the performance of the multiple sinusoidal model is quite good in analyzing variable star data.

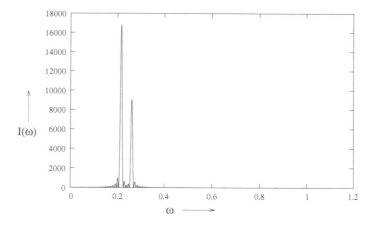

Fig. 6.4 The plot of the periodogram function of the variable star data

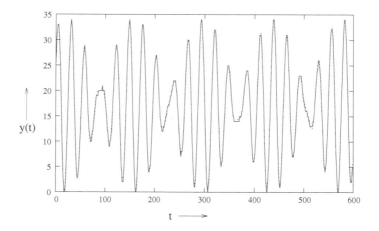

Fig. 6.5 The observed variable star data (*solid line*) along with the estimated values

6.4 "uuu" Data

In the "uuu" voiced speech data, 512 signal values, sampled at 10 kHz frequency, are available. The mean corrected data are displayed in Fig. 6.6. The plot in Fig. 6.6 suggests that the signal is mean non-stationary and there exists strong periodicity. The periodogram function of the "uuu" data is plotted in Fig. 6.7 and we obtain $\widehat{p} = 4$. The estimated parameters $(\widehat{A}_k, \widehat{B}_k, \widehat{\omega}_k), k = 1, \ldots, 4$ for "uuu" data are listed in Table 6.1. These point estimates are used in estimating the fitted/predicted signal. The predicted signal (solid line) of the mean corrected data along with the mean corrected observed "uuu" data (dotted line) are presented in Fig. 6.8. The fitted values match quite well with the mean corrected observed data.

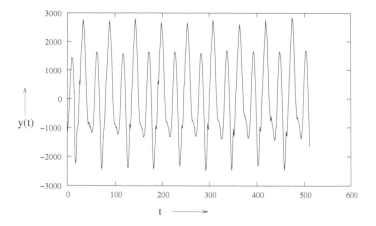

Fig. 6.6 The plot of the mean corrected "uuu" vowel sound data

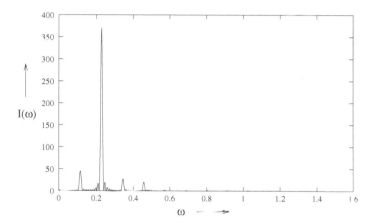

Fig. 6.7 The plot of the periodogram function of "uuu" vowel data

6.5 Airline Passenger Data

This is a classical data in time series analysis. The data represent the monthly international airline passenger data during January 1953 to December 1960, collected from the Time Series Data Library of Hyndman (n.d.). The raw data are plotted in Fig. 1.3. It is clear from the plot that the variability increases with time, so it cannot be considered as constant variance case. The log transform of the data is plotted in Fig. 6.9 to stabilize the variance. The variance seems to be approximately constant now, but at the same time there is a significant linear trend component present along with multiple periodic components. Therefore, a transformation of the form $a + bt$ or application of the difference operator to the log transform data is required. We choose

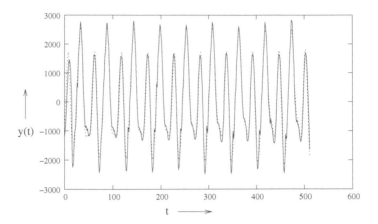

Fig. 6.8 The plot of the fitted values (*solid line*) and the mean corrected "uuu" vowel data

Table 6.1 The point estimates of the unknown parameters for variable star data and "uuu" data

Data Set: "Variable Star"					
A_1	7.48262	B_1	7.46288	ω_1	0.21623
A_2	−1.85116	B_2	6.75062	ω_2	0.26181
A_3	−0.80728	B_3	0.06880	ω_2	0.21360
Data Set: "uuu"					
A_1	8.92728	B_1	1698.65247	ω_1	0.22817
A_2	−584.67987	B_2	−263.79034	ω_2	0.11269
A_3	−341.40890	B_3	−282.07540	ω_3	0.34326
A_4	−193.93609	B_4	−300.50961	ω_4	0.45770

to use the difference operator and finally we have $y(t) = \log x(t+1) - \log x(t)$, which can be analyzed using multiple sinusoidal model. The transformed data are plotted in Fig. 6.10. It now appears that there is no trend component with approximately constant variance. Here, $\{x(t)\}$ represents the observed data. Now to estimate the frequency components and to get an idea about the number of frequency components present in $\{y(t)\}$, we plot the periodogram function of $\{y(t)\}$ in the interval $(0, \pi)$ in Fig. 6.11. This dataset has been analyzed by Nandi and Kundu [2].

There are six peaks corresponding to dominating frequency components in the periodogram plot. The initial estimates of $\omega_1, \ldots, \omega_6$ are obtained one by one using the sequential procedures. After taking out the effects of these six frequency components, we again study the periodogram function of the residual plot, similarly as in the case of the variable star data. We observe that there is an additional significant frequency component present. Hence, we estimate p as seven and accordingly estimate the other parameters. Finally plugging in the estimated parameters, we have the fitted series which is plotted in Fig. 6.12 along with the log difference data. They match quite well. The sum of squares of the residuals of this sinusoidal fit

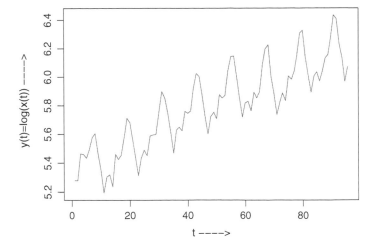

Fig. 6.9 The logarithm of the observed airline passenger data

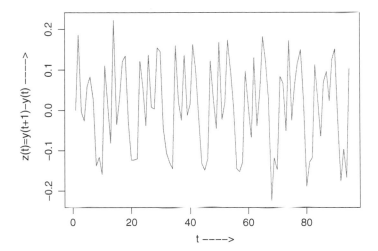

Fig. 6.10 The first difference of the log of the airline passenger data

is 5.54×10^{-4}. The monthly international airline passenger data is a well-studied dataset in time series literature. Usually a seasonal ARIMA (multiplicative) model is used to analyze it. A reasonable fit using this class of models is a seasonal ARIMA of order $(0, 1, 1) \times (0, 1, 1)_{12}$ to the log of the data, which is same as the model $(0, 0, 1) \times (0, 1, 1)_{12}$ applied to the difference of the log data (discussed by Box et al. [3] in detail). The fitted seasonal ARIMA model to the difference of the log data $\{y(t)\}$ is

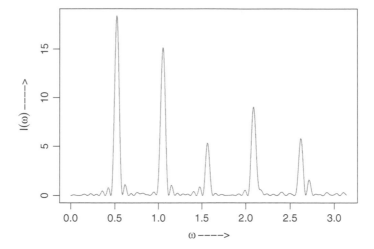

Fig. 6.11 The periodogram function of the log-difference data

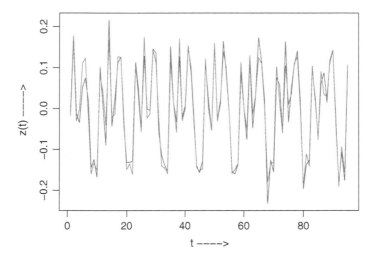

Fig. 6.12 The fitted values (*green*) along with the log difference data (*red*)

$$y(t) = y(t - 12) + Z(t) + 0.3577Z(t - 1) + 0.4419Z(t - 12) + 0.1581Z(t - 13),$$

where $\{Z(t)\}$ is a white noise sequence with mean zero and estimated variance
0.001052. In this case, we observe that the residual sum of squares is 9.19×10^{-4},
which is greater than the residual sum of squares of the proposed method.

6.6 Conclusions

Basic analysis of four data sets from different fields of application have been presented. It has been observed that the multiple sinusoidal model is effective to capture the periodicity present in these data. If p is large, as in the case of ECG signal data, the usual least squares method has difficulties in estimating such a large number (in the order of $n/2$) of unknown parameters. In such cases, the sequential method described in Sect. 3.12 is extremely useful. In all the data analyses considered here, the stationary error assumption is satisfied. The model is able to extract inherent periodicity, if present, from the transformed data also.

References

1. Prasad, A., Kundu, D., & Mitra, A. (2008). Sequential estimation of the sum of sinusoidal model parameters. *Journal of Statistical Planning and Inference, 138*, 1297–1313.
2. Nandi, S., & Kundu, D. (2011). Estimation of parameters of partially sinusoidal frequency model. *Statistics*, doi:10.1080/02331888.2011.577894.
3. Box, G. E. P., Jenkins, G. M., & Reinsel, G. C. (2008). *Time series analysis, forecasting and control* (4th ed.). New York: Wiley.

Chapter 7
Multidimensional Models

7.1 Introduction

In the last few chapters, we have discussed different aspects of 1-D sinusoidal frequency model. In this chapter, our aim is to introduce 2-D and 3-D frequency models and discuss several issues related to them. The 2-D and 3-D sinusoidal frequency models are natural generalizations of 1-D sinusoidal frequency model (3.1), and these models have various applications.

The 2-D sinusoidal frequency model has the following form;

$$y(m, n) = \sum_{k=1}^{p} [A_k \cos(m\lambda_k + n\mu_k) + B_k \sin(m\lambda_k + n\mu_k)] + X(m, n);$$

$$\text{for} \quad m = 1, \ldots, M, \quad n = 1, \ldots, N. \tag{7.1}$$

Here for $k = 1, \ldots, p$, A_k, B_k are unknown real numbers, λ_k, μ_k are unknown frequencies, and $\{X(m, n)\}$ is a 2-D sequence of error random variables with mean zero and finite variance. Several correlation structures have been assumed in the literature, and they are explicitly mentioned later. Two problems are of major interest associated to model (7.1). One is the estimation of A_k, B_k, λ_k, μ_k for $k = 1, \ldots, p$, and the other is the estimation of the number of components namely p. The first problem has received considerable attention in the statistical signal processing literature.

In the particular case when $\{X(m, n); m = 1, \ldots, M, n = 1, \ldots, N\}$ are i.i.d. random variables, this problem can be interpreted as a 'signal detection' problem, and it has different applications in 'Multidimensional Signal Processing'. This is a basic model in many fields such as antenna array processing, geophysical perception, biomedical spectral analysis, etc., see, for example, the work of Barbieri and Barone [1], Cabrera and Bose [2], Chun and Bose [3], Hua [4], and the references cited therein. This problem has a special interest in spectrography, and it has been studied using the group theoretic method by Malliavan [5, 6].

D. Kundu and S. Nandi, *Statistical Signal Processing*, SpringerBriefs in Statistics,
DOI: 10.1007/978-81-322-0628-6_7, © The Author(s) 2012

Fig. 7.1 The image plot of
a simulated data from model
(7.2)

Zhang and Mandrekar [7] used the 2-D sinusoidal frequency model for analyzing symmetric gray-scale textures. Any gray-scale picture, stored in digital format in a computer, is composed of tiny dots or pixels. In the digital representation of gray-scale or black and white pictures, the gray shade at each pixel is determined by the gray-scale intensity at that pixel. For instance, if 0 represents black and 1 represents white, then any real number $\in [0, 1]$ corresponds to a particular intensity of gray. If a picture is composed of 2-D array of M pixels arranged in rows and N pixels arranged in columns, the size of the picture is $M \times N$ pixels. The gray-scale intensities corresponding to the gray shades of various pixels can be stored in an $M \times N$ matrix. This transformation from a picture to a matrix and again back to a picture can be easily performed by image processing tools of any standard mathematical software.

Consider the following synthesized gray-scale texture data for $m = 1, \ldots, 40$ and $n = 1, \ldots, 40$.

$$y(m, n) = 4.0 \cos(1.8m + 1.1n) + 4.0 \sin(1.8m + 1.1n)$$
$$1.0 \cos(1.7m + 1.0n) + 1.0 \sin(1.7m + 1.0n) + X(m, n), \quad (7.2)$$

here $X(m, n) = e(m, n) + 0.25e(m - 1, n) + 0.25e(m, n - 1)$, and $e(m, n)$ for $m = 1, \ldots, 40$ and $n = 1, \ldots, 40$, are i.i.d. normal random variables with mean 0 and variance 2.0. This texture is displayed in Fig. 7.1. From Fig. 7.1 it is clear that model (7.1) can be used quite effectively to generate symmetric textures.

In this chapter, we briefly discuss about different estimators of the unknown parameters of 2-D model (7.1) and their properties. It can be seen that model (7.1) is a non-linear regression model, and therefore, the LSEs seem to be the most reasonable estimators. It is observed that the 1-D periodogram method can be extended to 2-D model, but it also has similar problems as the 1-D periodogram method. Although, the LSEs are the most intuitive natural estimators in this case, it is well known that finding the LSEs is a computationally challenging problem, particularly when p is

Fig. 7.2 The image plot of
a simulated data from model
(7.3)

large. It is observed that the sequential method recently proposed by Prasad et al.
[8] can be used very effectively in producing estimators, which are equivalent to the
LSEs. For $p = 1$, an efficient algorithm has been proposed by Nandi et al. [9] which
produces estimators which are equivalent to the LSEs in three steps starting from the
PEs.

Similarly, 3-D sinusoidal frequency model takes the following form;

$$
\begin{aligned}
y(m, n, s) = \sum_{k=1}^{p} [A_k &\cos(m\lambda_k + n\mu_k + sv_k) \\
&+ B_k \sin(m\lambda_k + n\mu_k + sv_k)] + X(m, n, s); \\
&\text{for } m = 1, \ldots, M, \quad n = 1, \ldots, N, \quad s = 1, \ldots, S. \quad (7.3)
\end{aligned}
$$

Here for $k = 1, \ldots, p$, A_k, B_k are unknown real numbers, λ_k, μ_k, v_k are unknown
frequencies, $\{X(m, n, s)\}$ is a 3-D sequence of error random variables with mean
zero, finite variance, and 'p' denotes the number of 3-D sinusoidal components.
Model (7.3) has been used for describing color textures by Prasad and Kundu [10],
see Fig. 7.2. The third dimension represents different color schemes. In digital representation, any color picture is stored digitally in RGB format. Almost any color can
be represented by a unique combination of red, green, and blue color intensities. In
RGB format, $S = 3$. A color picture can be stored digitally in an $M \times N \times S$ array.
Similarly as a black and white picture, any image processing tool of a mathematical software can convert a color picture to a 3-D array and vice versa. For detailed
description on how a color picture is stored digitally, the readers are referred to Prasad
and Kundu [10].

7.2 2-D Model: Estimation of Frequencies

7.2.1 LSEs

As it has been mentioned earlier that in the presence of i.i.d. errors, the LSEs seem to be the most natural choice, and they can be obtained by minimizing

$$\sum_{m=1}^{M}\sum_{n=1}^{N}\left(y(m,n)-\sum_{k=1}^{p}[A_k\cos(m\lambda_k+n\mu_k)+B_k\sin(m\lambda_k+n\mu_k)]\right)^2, \quad (7.4)$$

with respect to the unknown parameters. Minimization of (7.4) can be performed in two steps by using the separable regression technique of Richards [11]. For $k=1,\ldots,p$, when λ_k and μ_k are fixed, the LSEs of A_k and B_k can be obtained as

$$\left[\widehat{A}_1(\lambda,\mu),\widehat{B}_1(\lambda,\mu):\ldots:\widehat{A}_p(\lambda,\mu),\widehat{B}_p(\lambda,\mu)\right]^T=\left(\mathbf{U}^T\mathbf{U}\right)^{-1}\mathbf{U}^T\mathbf{Y}, \quad (7.5)$$

here $\lambda=(\lambda_1,\ldots,\lambda_p)$, $\mu=(\mu_1,\ldots,\mu_p)$, $\mathbf{U}$ is an $MN\times 2p$ matrix and $\mathbf{Y}$ is an $MN\times 1$ vector as follows;

$$\mathbf{U}=[\mathbf{U}_1:\cdots:\mathbf{U}_p]$$

$$\mathbf{U}_k=\begin{bmatrix}\cos(\lambda_k+\mu_k)&\cdots&\cos(\lambda_k+N\mu_k)&\cos(2\lambda_k+\mu_k)&\cdots&\cos(M\lambda_k+N\mu_k)\\\sin(\lambda_k+\mu_k)&\cdots&\sin(\lambda_k+N\mu_k)&\sin(2\lambda_k+\mu_k)&\cdots&\sin(M\lambda_k+N\mu_k)\end{bmatrix}^T,$$

$$(7.6)$$

for $k=1,\ldots,p$, and

$$\mathbf{Y}=\left(y(1,1)\cdots y(1,N)\,y(2,1)\cdots y(2,N)\cdots y(M,1)\cdots y(M,N)\right)^T. \quad (7.7)$$

For $k=1,\ldots,p$, once $\widehat{A}_k(\lambda,\mu)$ and $\widehat{B}_k(\lambda,\mu)$ are obtained, the LSEs of λ_k and μ_k are obtained by minimizing

$$\sum_{m=1}^{M}\sum_{n=1}^{N}\left(y(m,n)-\sum_{k=1}^{p}\left[\widehat{A}_k(\lambda,\mu)\cos(m\lambda_k+n\mu_k)+\widehat{B}_k(\lambda,\mu)\sin(m\lambda_k+n\mu_k)\right]\right)^2,$$

$$(7.8)$$

with respect to $\lambda_1,\ldots,\lambda_p$ and $\mu_1,\ldots,\mu_p$. Once $\widehat{\lambda}_k$ and $\widehat{\mu}_k$, the LSEs of λ_k and μ_k respectively, are obtained, the LSEs of A_k and B_k can be obtained as $\widehat{A}_k(\widehat{\lambda},\widehat{\mu})$ and $\widehat{B}_k(\widehat{\lambda},\widehat{\mu})$, respectively, where $\widehat{\lambda}=(\widehat{\lambda}_1,\ldots,\widehat{\lambda}_p)$ and $\widehat{\mu}=(\widehat{\mu}_1,\ldots,\widehat{\mu}_p)$. The minimization of (7.8) can be obtained by solving a $2p$-dimensional optimization problem, which can be computationally quite challenging if p is large.

Rao et al. [12] obtained the theoretical properties of the LSEs of the parameters of a similar model namely 2-D superimposed complex exponential model, that is,

$$y(m, n) = \sum_{k=1}^{p} C_k e^{i(m\lambda_k + n\mu_k)} + Z(m, n),\tag{7.9}$$

here for $k = 1, \ldots, p$, C_k are complex numbers, λ_k and μ_k are same as defined before, $\{Z(m, n)\}$ is a 2-D sequence of complex-valued random variables. Rao et al. [12] obtained the consistency and asymptotic normality properties of the LSEs under the assumptions that $\{Z(m, n); m = 1, \ldots, M, n = 1, \ldots, N\}$ are i.i.d. complex normal random variables. Kundu and Gupta [13] proved the consistency and asymptotic normality properties of the LSEs of model (7.1) under the assumption that the error random variables $\{X(m, n); m = 1, \ldots, M, n = 1, \ldots, N\}$ are i.i.d. with mean zero and finite variance. Later Kundu and Nandi [14] provided the consistency and asymptotic normality properties of the LSEs under the following stationary assumptions on $\{X(m, n)\}$.

Assumption 7.1 The double array sequence of random variables $\{X(m, n)\}$ can be represented as follows;

$$X(m, n) = \sum_{j=-\infty}^{\infty} \sum_{k=-\infty}^{\infty} a(j, k) e(m - j, n - k),$$

where $a(j, k)$ are real constants such that

$$\sum_{j=-\infty}^{\infty} \sum_{k=-\infty}^{\infty} |a(j, k)| < \infty,$$

and $\{e(m, n)\}$ is a double array sequence of i.i.d. random variables with mean zero and variance σ^2.

We use the following notation to provide the asymptotic distribution of the LSEs of the parameters of model (7.1) obtained by Kundu and Nandi [14]:

$$\boldsymbol{\theta}_1 = (A_1, B_1, \lambda_1, \mu_1), \ldots, \boldsymbol{\theta}_p = (A_p, B_p, \lambda_p, \mu_p)$$

$$\mathbf{D} = \mathrm{diag}\{M^{1/2}N^{1/2}, M^{1/2}N^{1/2}, M^{3/2}N^{1/2}, M^{1/2}N^{3/2}\},$$

and for $k = 1, \ldots, p$,

$$
\Sigma_k =
\begin{bmatrix}
1 & 0 & \frac{1}{2}B_k & \frac{1}{2}B_k \\[2mm]
0 & 1 & -\frac{1}{2}A_k & -\frac{1}{2}A_k \\[2mm]
\frac{1}{2}B_k & -\frac{1}{2}A_k & \frac{1}{3}(A_k^2 + B_k^2) & \frac{1}{4}(A_k + B_k) \\[2mm]
\frac{1}{2}B_k & -\frac{1}{2}A_k & \frac{1}{4}(A_k + B_k) & \frac{1}{3}(A_k^2 + B_k^2)
\end{bmatrix}.
$$

Theorem 7.1 *Under Assumption 7.1, $\widehat{\boldsymbol{\theta}}_1, \ldots, \widehat{\boldsymbol{\theta}}_p$, the LSEs of $\boldsymbol{\theta}_1, \ldots, \boldsymbol{\theta}_p$ respectively, are consistent estimators, and as $\min\{M, N\} \to \infty$*

$$
\left(\mathbf{D}(\widehat{\boldsymbol{\theta}}_1 - \boldsymbol{\theta}_1), \ldots, \mathbf{D}(\widehat{\boldsymbol{\theta}}_p - \boldsymbol{\theta}_p)\right) \xrightarrow{d} \mathcal{N}_{4p}\left(\mathbf{0}, 2\sigma^2 \Delta^{-1}\right),
$$

here

$$
\Delta^{-1} =
\begin{bmatrix}
c_1 \Sigma_1^{-1} & 0 & \cdots & 0 \\
0 & c_2 \Sigma_2^{-1} & \cdots & 0 \\
\vdots & & \ddots & \vdots & \vdots \\
0 & 0 & \cdots & c_p \Sigma_p^{-1}
\end{bmatrix}
$$

and for $k = 1, \ldots, p$,

$$
c_k = \left| \sum_{u=-\infty}^{\infty} \sum_{v=-\infty}^{\infty} a(u, v) e^{-i(u\lambda_k + v\mu_k)} \right|^2.
$$

From Theorem 7.1, it is clear that even for 2-D model, the LSEs of the frequencies have much faster convergence rates than the linear parameters.

7.2.2 Sequential Method

It has been observed that the LSEs are the most efficient estimators, although, computing the LSEs is a difficult problem. As it has been mentioned in the previous section that it involves solving a $2p$-dimensional optimization problem. It might be quite difficult particularly if p is large. To avoid this problem Prasad et al. [8] proposed a sequential estimation procedure of the unknown parameters, which have the same rate of convergence as the LSEs. Moreover, the sequential estimators can be obtained by solving p, 2-D optimization problems sequentially. Therefore, even if p is large, sequential estimators can be obtained quite easily compared to the LSEs. It can be described as follows. At the first step, minimize the quantity

$$Q_1(A, B, \lambda, \mu) = \sum_{m=1}^{M} \sum_{n=1}^{N} (y(m, n) - A\cos(m\lambda + n\mu) - B\sin(m\lambda + n\mu))^2,$$

(7.10)

with respect to A, B, λ, μ. It can be easily seen using the separable regression technique of Richards [11] that for fixed λ and μ, $\widetilde{A}(\lambda, \mu)$ and $\widetilde{B}(\lambda, \mu)$ minimize (7.10), where

$$\left[\widetilde{A}(\lambda, \mu)\ \widetilde{B}(\lambda, \mu) \right]^T = \left(\mathbf{U}_1^T \mathbf{U}_1 \right)^{-1} \mathbf{U}_1^T \mathbf{Y},$$

(7.11)

where $\mathbf{U}_1$ is an $MN \times 2$ and $\mathbf{Y}$ is an $MN \times 1$ data vector as defined in (7.6) and (7.7), respectively. Replacing $\widetilde{A}(\lambda, \mu)$ and $\widetilde{B}(\lambda, \mu)$ in (7.10), we obtain

$$R_1(\lambda, \mu) = Q_1(\widetilde{A}(\lambda, \mu), \widetilde{B}(\lambda, \mu), \lambda, \mu).$$

(7.12)

If $\widetilde{\lambda}$ and $\widetilde{\mu}$ minimize $R_1(\lambda, \mu)$, $\left(\widetilde{A}(\widetilde{\lambda}, \widetilde{\mu}), \widetilde{B}(\widetilde{\lambda}, \widetilde{\mu}), \widetilde{\lambda}, \widetilde{\mu} \right)$ minimizes (7.10). Denote these estimators as $\widetilde{\boldsymbol{\theta}}_1 = \left(\widetilde{A}_1, \widetilde{B}_1, \widetilde{\lambda}_1, \widetilde{\mu}_1 \right)$.

Consider $\{y_1(m, n); m = 1, \ldots, M, n = 1, \ldots, N\}$, where

$$y_1(m, n) = y(m, n) - \widetilde{A}_1\cos(m\widetilde{\lambda}_1 + n\widetilde{\mu}_1) - \widetilde{B}_1\sin(m\widetilde{\lambda}_1 + n\widetilde{\mu}_1).$$

(7.13)

Repeating the whole procedure as described above by replacing $y(m, n)$ with $y_1(m, n)$, obtain $\widetilde{\boldsymbol{\theta}}_2 = \left(\widetilde{A}_2, \widetilde{B}_2, \widetilde{\lambda}_2, \widetilde{\mu}_2 \right)$. Following along the same line, one can obtain $\widetilde{\theta}_1, \ldots, \widetilde{\theta}_p$. Prasad et al. [8] proved that Theorem 7.1 also holds for $\widetilde{\theta}_1, \ldots, \widetilde{\theta}_p$. Further $\widetilde{\theta}_j$ and $\widetilde{\theta}_k$, $j \neq k$ are independently distributed. It implies that the LSEs and the estimators obtained by using the sequential method are asymptotically equivalent estimators.

7.2.3 Periodogram Estimators

The 2-D periodogram function of any double array sequence of observations $\{y(m, n); m = 1, \ldots, M, n = 1, \ldots, N\}$ is defined as follows;

$$I(\lambda, \mu) = \frac{1}{MN} \left| \sum_{m=1}^{M} \sum_{n=1}^{N} y(m, n)e^{-i(m\lambda + n\mu)} \right|^2.$$

(7.14)

Here, the 2-D periodogram function is evaluated at the 2-D Fourier frequencies, namely at $(\pi k/M, \pi j/N); k = 0, \ldots, M, j = 0, \ldots, N$. Clearly the 2-D periodogram function (7.14) is a natural generalization of 1-D periodogram function. Zhang and Mandrekar [7] and Kundu and Nandi [14] used the periodogram function (7.14) to estimate the frequencies of model (7.1). For $p = 1$, Kundu and Nandi [14] proposed the ALSEs as follows;

$$(\widetilde{\lambda}, \widetilde{\mu}) = \arg \max I(\lambda, \mu), \qquad (7.15)$$

where the maximization is performed for $0 \le \lambda \le \pi$ and $0 \le \mu \le \pi$, and the estimates of A and B are obtained as follows;

$$\widetilde{A} = \frac{2}{MN} \sum_{m=1}^{M} \sum_{m=1}^{N} y(m, n) \cos(m\widetilde{\lambda} + n\widetilde{\mu}) \qquad (7.16)$$

$$\widetilde{B} = \frac{2}{MN} \sum_{m=1}^{M} \sum_{m=1}^{N} y(m, n) \sin(m\widetilde{\lambda} + n\widetilde{\mu}). \qquad (7.17)$$

It has been shown that under Assumption 7.1, the asymptotic distribution of $(\widetilde{A}, \widetilde{B}, \widetilde{\lambda}, \widetilde{\mu})$ is same as in Theorem 7.1. Sequential method as described in the previous section can be applied exactly in the same manner for general p, and the asymptotic distribution of the ALSEs also satisfies Theorem 7.1. Therefore, the LSEs and ALSEs are asymptotically equivalent.

7.2.4 Nandi–Prasad–Kundu Algorithm

The ALSEs or the sequential estimators as described in the previous two sections can be obtained by solving a 2-D optimization problem. It is well known that the least squares surface and the periodogram surface have several local minima and local maxima respectively. Therefore, the convergence of any optimization algorithm is not guaranteed. Recently Nandi et al. [9] proposed a three-step algorithm which produces estimators of the unknown frequencies which have the same rate of convergence as the LSEs. We provide the algorithm for $p = 1$. The sequential procedure can be easily used for general p. We use the following notation for describing the algorithm.

$$P_{MN}^1(\lambda, \mu) = \sum_{t=1}^{M} \sum_{s=1}^{N} \left(t - \frac{M}{2} \right) y(t, s) e^{-i(\lambda t + \mu s)} \qquad (7.18)$$

$$P_{MN}^2(\lambda, \mu) = \sum_{t=1}^{M} \sum_{s=1}^{N} \left(s - \frac{N}{2} \right) y(t, s) e^{-i(\lambda t + \mu s)} \qquad (7.19)$$

$$Q_{MN}(\lambda, \mu) = \sum_{t=1}^{M} \sum_{s=1}^{N} y(t, s) e^{-i(\lambda t + \mu s)} \qquad (7.20)$$

$$\widehat{\lambda}^{(r)} = \widehat{\lambda}^{(r-1)} + \frac{12}{M_r^2} \text{Im} \left[\frac{P_{M_r N_r}^1(\widehat{\lambda}^{(r-1)}, \widehat{\mu}^{(0)})}{Q_{M_r N_r}(\widehat{\lambda}^{(r-1)}, \widehat{\mu}^{(0)})} \right], \quad r = 1, 2, \ldots, \qquad (7.21)$$

$$\widehat{\mu}^{(r)} = \widehat{\mu}^{(r-1)} + \frac{12}{N_r^2} \text{Im} \left[\frac{P_{M_r N_r}^2(\widehat{\lambda}^{(0)}, \widehat{\mu}^{(r-1)})}{Q_{M_r N_r}(\widehat{\lambda}^{(0)}, \widehat{\mu}^{(r-1)})} \right], \quad r = 1, 2, \ldots. \qquad (7.22)$$

Nandi et al. [9] suggested to use the following initial guesses of λ and μ. For any fixed $n \in \{1, \ldots, N\}$ from the data vector $\{y(1, n), \ldots, y(M, n)\}$, obtain the periodogram maximizer over Fourier frequencies, say at $\widehat{\lambda}_n$. Take

$$\widehat{\lambda}^{(0)} = \frac{1}{N} \sum_{n=1}^{N} \widehat{\lambda}_n. \tag{7.23}$$

Similarly, for fixed $m \in \{1, \ldots, M\}$, from the data vector $\{y(m, 1), \ldots, y(m, N)\}$, first obtain $\widehat{\mu}_m$, which is the periodogram maximizer over Fourier frequencies, and then consider

$$\widehat{\mu}^{(0)} = \frac{1}{M} \sum_{m=1}^{M} \widehat{\mu}_m. \tag{7.24}$$

It has been shown that $\widehat{\lambda}^{(0)} = O_p(M^{-1}N^{-1/2})$ and $\widehat{\mu}^{(0)} = O_p(M^{-1/2}N^{-1})$. The algorithm can be described as follows.

Algorithm 7.1

- Step 1: Take $r = 1$, choose $M_1 = M^{0.8}$, $N_1 = N$. Compute $\widehat{\lambda}^{(1)}$ from $\widehat{\lambda}^{(0)}$ using (7.21).
- Step 2: Take $r = 2$, choose $M_2 = M^{0.9}$, $N_2 = N$. Compute $\widehat{\lambda}^{(2)}$ from $\widehat{\lambda}^{(1)}$ using (7.21).
- Step 3: Take $r = 3$, choose $M_3 = M$, $N_3 = N$. Compute $\widehat{\lambda}^{(3)}$ from $\widehat{\lambda}^{(2)}$ using (7.21).

Exactly in the same manner $\widehat{\mu}^{(3)}$ can also be obtained from (7.22). $\widehat{\lambda}^{(3)}$ and $\widehat{\mu}^{(3)}$ are the proposed estimators of λ and μ respectively. It has been shown that the proposed estimators have the same asymptotic variances as the corresponding LSEs. The main advantage of the proposed estimators is that they can be obtained in a fixed number of iterations. Extensive simulation results suggest that the proposed algorithm works very well.

7.2.5 Noise Space Decomposition Method

Recently Nandi et al. [15] proposed the noise space decomposition method to estimate the frequencies of the 2-D sinusoidal model (7.1). The proposed method is an extension of 1-D NSD method which was originally proposed by Kundu and Mitra [16] as described in Sect. 3.4. The NSD method for 2-D model can be described as follows: From the sth row of the data matrix

$$\mathbf{Y}_N = \begin{bmatrix} y(1,1) & \cdots & y(1,N) \\ \vdots & \ddots & \vdots \\ y(s,1) & \cdots & y(s,N) \\ \vdots & \ddots & \vdots \\ y(M,1) & \cdots & y(M,N) \end{bmatrix}, \tag{7.25}$$

construct a matrix $\mathbf{A}_s$ for any $N - 2p \ge L \ge 2p$ as follows,

$$\mathbf{A}_s = \begin{bmatrix} y(s,1) & \cdots & y(s,L+1) \\ \vdots & \vdots & \vdots \\ y(s,N-L) & \cdots & y(s,N) \end{bmatrix}.$$

Obtain an $(L+1) \times (L+1)$ matrix $\mathbf{B} = \sum_{s=1}^{M} \mathbf{A}_s^T \mathbf{A}_s / ((N-L)M)$. Now using the 1-D NSD method on matrix $\mathbf{B}$, the estimates of $\lambda_1, \ldots, \lambda_p$ can be obtained. Similarly, using the columns of the data matrix $\mathbf{Y}_N$, the estimates of $\mu_1, \ldots, \mu_p$ can be obtained. For details see Nandi et al. [15]. Finally one needs to estimate the pairs, namely $\{(\lambda_k, \mu_k); k = 1, \ldots, p\}$ also. The authors suggested the following two pairing algorithms once the estimates of λ_k and μ_k for $k = 1, \ldots, p$ are obtained. Algorithm 7.2 is based on $p!$ search. It is computationally efficient for small values of p, say $p = 2, 3$ and Algorithm 7.3 is based on p^2-search, so it is efficient for large values of p, that is, when p is greater than 3. Suppose the estimates obtained using the above NSD method are $\{\widehat{\lambda}_{(1)}, \ldots, \widehat{\lambda}_{(p)}\}$ and $\{\widehat{\mu}_{(1)}, \ldots, \widehat{\mu}_{(p)}\}$, then the two algorithms are described as follows.

Algorithm 7.2 Consider all possible $p!$ combinations of pairs $\{(\widehat{\lambda}_{(j)}, \widehat{\mu}_{(j)}) : j = 1, \ldots, p\}$ and calculate the sum of the periodogram function for each combination as

$$I_S(\boldsymbol{\lambda}, \boldsymbol{\mu}) = \sum_{k=1}^{p} \frac{1}{MN} \left| \sum_{s=1}^{M} \sum_{t=1}^{N} y(s,t) e^{-i(s\lambda_k + t\mu_k)} \right|^2.$$

Consider that combination as the paired estimates of $\{(\lambda_j, \mu_j) : j = 1, \ldots, p\}$ for which this $I_S(\boldsymbol{\lambda}, \boldsymbol{\mu})$ is maximum.

Algorithm 7.3 Compute $I(\lambda, \mu)$ as defined in (7.14) over $\{(\widehat{\lambda}_{(j)}, \widehat{\mu}_{(k)}), j, k = 1, \ldots, p\}$. Choose the largest p values of $I(\widehat{\lambda}_{(j)}, \widehat{\mu}_{(k)})$ and the corresponding $\{(\widehat{\lambda}_{[k]}, \widehat{\mu}_{[k]}), k = 1, \ldots, p\}$ are the paired estimates of $\{(\lambda_k, \mu_k), k = 1, \ldots, p\}$.

From the extensive experimental results it is observed that the performance of these estimators is better than that of the ALSEs, and compare reasonably well with the LSEs. It has been observed along the same line as the 1-D NSD method that under the assumptions of i.i.d. errors, the frequency estimators obtained by the 2-D

NSD method are strongly consistent, although the asymptotic distribution of these estimators has not yet been established.

7.3 2-D Model: Estimating the Number of Components

The estimation of frequencies of the 2-D sinusoidal model has received considerable attention in the signal processing literature. Unfortunately, not that much of attention has been paid in estimating the number of components namely p of model (7.1). It may be mentioned that p can be estimated by observing the number of peaks of the 2-D periodogram function, as mentioned in Kay [17], but that is quite subjective in nature and it may not work properly all the times.

Miao et al. [18] discussed the estimation of the number of components for an equivalent model. Kundu and Nandi [14] proposed a method based on the eigen-decomposition technique and it avoids estimation of the different parameters for different model orders. It only needs the estimation of error variance for different model orders. The method uses the rank of a Vandermond-type matrix and the information theoretic criteria like AIC and MDL. But instead of using any fixed penalty function a class of penalty functions satisfying some special properties has been used. It is observed that any penalty function from that particular class provides consistent estimates of the unknown parameter p under the assumptions that the errors are i.i.d. random variables. Further, an estimate of probability of wrong detection for any particular penalty function has been obtained using the matrix perturbation technique. Once an estimate of the probability of wrong detection has been obtained, that penalty function from the class of penalty functions for which the estimated probability of wrong detection is minimum, is used to estimate p. The main feature of this method is that the penalty function depends on the observed data, and it has been observed by extensive numerical studies that the data-dependent penalty function works very well in estimating p.

7.4 Conclusions

In this chapter, we have considered the 2-D sinusoidal model and discussed different estimators and their properties. In the 2-D case also, the LSEs are the most natural choice and they are the most efficient estimators also. Unfortunately, finding the LSEs is a difficult problem. Due to this reason, several other estimators which may not be as efficient as the LSEs, but easier to compute have been suggested in the literature. Recently Prasad and Kundu [10] proposed a 3-D sinusoidal model, which can be used quite effectively to model color textures. They have established the strong consistency and asymptotic normality properties of the LSEs under the assumptions of stationary errors. No attempt has been made to compute the LSEs efficiently or to find some other estimators which can be obtained more conveniently than the LSEs.

It seems most of the 2-D results should be possible to extend to 3-D case. More work is needed along that direction.

References

1. Barbieri, M. M., & Barone, P. (1992). A two-dimensional Prony's algorithm for spectral estimation. *IEEE Transaction on Signal Processing, 40,* 2747–2756.
2. Cabrera, S. D. & Bose, N. K. (1993). Prony's method for two-dimensional complex exponential modeling (Chapter 15). In S. G. Tzafestas (Ed.), *Applied Control Theory* (pp. 401–411). New York, NY: Marcel and Dekker.
3. Chun, J., & Bose, N. K. (1995). Parameter estimation via signal selectivily of signal subspaces (PESS) and its applications. *Digital Signal Processing, 5,* 58–76.
4. Hua, Y. (1992). Estimating two-dimensional frequencies by matrix enhancement and matrix Pencil. *IEEE Transaction on Signal Processing, 40,* 2267–2280.
5. Malliavan, P. (1994). Sur la norme d'une matrice circulate Gaussienne Serie I. *C.R. Acad. Sc. Paris t, 319,* 45–49.
6. Malliavan, P. (1994). Estimation d'un signal Lorentzien Serie I. *C.R. Acad. Sc. Paris t, 319,* 991–997.
7. Zhang, H., & Mandrekar, V. (2001). Estimation of hidden frequencies for 2D stationary processes. *Journal of Time Series Analysis, 22,* 613–629.
8. Prasad, A., Kundu, D. & Mitra, A. (2011). Sequential estimation of two dimensional sinusoidal models. *Journal of Probability and Statistical Science,* (to appear).
9. Nandi, S., Prasad, A., & Kundu, D. (2010). An efficient and fast algorithm for estimating the parameters of two-dimensional sinusodial signals. *Jorunal of Statistical Planning and Inference, 140,* 153–168.
10. Prasad, A., & Kundu, D. (2009). Modeling and estimation of symmetric color textures. *Sankhya Series B, 71,* 30–54.
11. Richards, F. S. G. (1961). A method of maximum likelihood estimation, *Journal of Royal Statistical Society, B23,* 469–475.
12. Rao, C. R., Zhao, L. C., & Zhou, B. (1994). Maximum likelihood estimation of 2-D superimposed exponential signals. *IEEE Transaction on Signal Processing, 42,* 1795–1802.
13. Kundu, D., & Gupta, R. D. (1998). Asymptotic properties of the least squares estimators of a two dimensional model. *Metrika, 48,* 83–97.
14. Kundu, D., & Nandi, S. (2003). Determination of discrete spectrum in a random field. *Statistica Neerlandica, 57,* 258–283.
15. Nandi, S., Kundu, D., & Srivastava, R. K. (2011). Noise space decomposition method for two-dimensional sinusoidal model. *Computational Statistics and Data Analysis,* (to appear).
16. Kundu, D., & Mitra, A. (1995). Consistent method of estimating the superimposed exponential signals. *Scandinavian Journal of Statistics, 22,* 73–82.
17. Kay, S. M. (1988). *Modern spectral estimation: Theory and application.* New York: Prentice Hall.
18. Miao, B. Q., Wu, Y., & Zhao, L. C. (1998). On strong consistency of a 2-dimensional frequency estimation algorithm. *Statistica Sinica, 8,* 559–570.

Chapter 8
Related Models

8.1 Introduction

The sinusoidal frequency model is a well known model in different fields of science and technology and as has been observed in previous chapters, is a very useful model in explaining nearly periodical data. There are several other models which are practically the multiple sinusoidal model, but also exploit some extra features in the data. In most of such cases, the parameters satisfy some additional conditions other than the assumptions required for the sinusoidal model. For example, if the frequencies appear at $\lambda, 2\lambda, \ldots, p\lambda$ in a multiple sinusoidal model, then the model that exploits this extra information is the *fundamental frequency model*. The advantage of using this information in the model itself is that it reduces the total number of parameters to $2p + 1$ from $3p$ and a single non-linear parameter instead of p, in case of multiple sinusoidal model. Similarly, if the gap between two consecutive frequencies is approximately same, then the suitable model is the *generalized fundamental frequency model*. We call these models as "related models" of the sinusoidal frequency model.

This chapter is organized in the following way. The damped sinusoid and the amplitude modulated (AM) model are discussed in Sects. 8.2 and 8.3, respectively. These are complex-valued models. The rest of the models, discussed here, are real-valued. The fundamental frequency model and the generalized fundamental frequency model are given in Sects. 8.4 and 8.5, respectively. The partial sinusoidal model is given in Sect. 8.6 and the burst model in Sect. 8.7. The chapter is concluded by a brief discussion of some more related models in Sect. 8.8.

8.2 Damped Sinusoidal Model

The superimposed damped exponential signal in the presence of noise is an important model in signal processing literature. It is a complex-valued model in general form and can be written as

D. Kundu and S. Nandi, *Statistical Signal Processing*, SpringerBriefs in Statistics,   113
DOI: 10.1007/978-81-322-0628-6_8, © The Author(s) 2012

$$y(t) = \mu(\boldsymbol{\alpha}, \boldsymbol{\beta}, t) + \varepsilon(t) = \sum_{j=1}^{p} \alpha_j \exp\{\beta_j t\} + \varepsilon(t), \quad t = t_i, \tag{8.1}$$

where t_i, $i = 1, \ldots, n$, are equidistant; α_j $j = 1, \ldots, p$ are unknown complex amplitudes; p is the total number of sinusoids present; β_j, $j = 1, \ldots, p$ are assumed to be distinct; $\{\varepsilon(t_1), \ldots, \varepsilon(t_n)\}$ are complex-valued random variables with mean zero and finite variance.

Model (8.1) represents a general complex-valued sinusoidal model and has three special cases; (i) undamped sinusoidal model, (ii) damped sinusoidal model, and (iii) real compartment model, depending on the form of β_j, $j = 1, \ldots, p$. When $\beta_j = i\omega_j$, $\omega_j \in (0, \pi)$, model (8.1) is an undamped sinusoid; if $\beta_j = -\delta_j + i\omega_j$ with $\delta_j > 0$ and $\omega_j \in (0, \pi)$ for all j, it is a damped sinusoid where δ_j is the damping factor and ω_j is the frequency corresponding to the jth component; if for all j, α_j and β_j are real numbers, model (8.1) represents a real compartment model. All the three models are quite common among engineers and scientists. For applications of damped and undamped models, the readers are referred to Kay [1] and for the real compartment model, see Seber and Wild [2] and Bates and Watts [3].

Tufts and Kumaresan consider model (8.1) with $t_i = i$. Some modifications of Prony's method were suggested by a series of authors. See Kumaresan [4], Kumaresan and Tufts [5], and the references therein. It is pointed out by Rao [6] that solutions obtained by these methods may not be consistent. Moreover, Wu [7] showed that any estimator of the unknown parameters of model (8.1) is inconsistent with $t_i = i$. Due to this reason, Kundu [8] considers the following alternative model. Write model (8.1) as

$$y_{ni} = \mu(\boldsymbol{\alpha}^0, \boldsymbol{\beta}^0, t_{ni}) + \varepsilon_{ni}, \quad i = 1, \ldots, n, \tag{8.2}$$

where $t_{ni} = i/n$, $i = 1, \ldots, n$ take values in the unit interval; $\boldsymbol{\theta} = (\boldsymbol{\alpha}, \boldsymbol{\beta}) = (\alpha_1, \ldots, \alpha_p, \beta_1, \ldots, \beta_p)$ be the parameter vector. Least norm squares estimator is the most natural estimator in this case. Kundu [9] extends the results of Jennrich [10] for the LSEs to be consistent and asymptotically normal to the complex parameter case. But the damped sinusoid does not satisfy Kundu's condition. It is necessary for the LSEs to be consistent that t_{ni}, $i = 1, \ldots, n$, $n = 1, 2, \ldots$ are bounded. It is also assumed that $\{\varepsilon_{ni}\}$, $i = 1, \ldots, n$, $n = 1, 2, \ldots$ is a double array sequence of complex-valued random variables. Each row $\{\varepsilon_{n1}, \ldots, \varepsilon_{nn}\}$ is i.i.d. with mean zero. The real and imaginary parts of ε_{ni} are independently distributed with finite fourth moments. The parameter space Θ is a compact subset of $\mathcal{C}^p$ and the true parameter vector $\boldsymbol{\theta}^0$ is an interior point of Θ. Further, the function

$$\int_0^1 |\mu(\boldsymbol{\alpha}^0, \boldsymbol{\beta}^0, t) - \mu(\boldsymbol{\alpha}, \boldsymbol{\beta}, t)|^2 dt \tag{8.3}$$

has a unique minimum at $(\boldsymbol{\alpha}, \boldsymbol{\beta}) = (\boldsymbol{\alpha}^0, \boldsymbol{\beta}^0)$. Under these assumptions, the LSEs of the unknown parameters of model (8.1) are strongly consistent.

8.3 Amplitude Modulated Model

This is a special type of AM undamped signal model and naturally complex-valued like damped or undamped model. The discrete-time complex random process $\{y(t)\}$ consisting of p single-tone AM signals is given by

$$y(t) = \sum_{k=1}^{p} A_k \left[1 + \mu_k e^{i \nu_k t} \right] e^{i \omega_k t} + X(t); \quad t = 1, \ldots, n, \qquad (8.4)$$

where for $k = 1, \ldots, p$, A_k is the carrier amplitude of the constituent signal, μ_k is the modulation index, ω_k is the carrier angular frequency, and ν_k is the modulating angular frequency. The sequence of additive errors $\{X(t)\}$ is a complex-valued stationary linear process.

The model was first proposed by Sircar and Syali [11]. Nandi and Kundu [12] and Nandi et al. [13] proposed LSEs and ALSEs and studied their theoretical properties for large n. The method, proposed by Sircar and Syali [11], is based on accumulated correlation functions, power spectrum, and Prony's difference-type equations, but applicable if $\{X(t)\}$ is a sequence of complex-valued i.i.d. random variables. Readers are referred to Sircar and Syali [11] for physical interpretation of different model parameters. Model (8.4) was introduced for analyzing some special type of non-stationary signal in steady-state analysis. If $\mu_k = 0$ for all k, model (8.4) coincides with the sum of complex exponential, that is, the undamped model. The undamped model is best suited for transient non-stationary signal, but it may lead to large order when the signal is not decaying over time. Sircar and Syali [11] argued that complex AM model is more suitable for steady-state non-stationary signal. This model was proposed to analyze some short duration speech data. Nandi and Kundu [12] and Nandi et al. [13] analyzed two such datasets.

The following restrictions are required on the true values of the model parameters: for all k, $A_k \neq 0$ and $\mu_k \neq 0$ and they are bounded. Also $0 < \nu_k < \pi$, $0 < \omega_k < \pi$ and

$$\omega_1 < \omega_1 + \nu_1 < \omega_2 < \omega_2 + \nu_2 < \cdots < \omega_M < \omega_M + \nu_M. \qquad (8.5)$$

A complex-valued stationary linear process implies that $X(t)$ has the following representation

$$X(t) = \sum_{k=0}^{\infty} a(k) e(t - k),$$

where $\{e(t)\}$ is a sequence of i.i.d. complex-valued random variables with mean zero and variance $\sigma^2 < \infty$ for both the real and imaginary parts. The real and imaginary parts of $e(t)$ are uncorrelated. The sequence $\{a(k)\}$ of arbitrary complex-valued constants is such that

$$\sum_{k=0}^{\infty} |a(k)| < \infty.$$

The norm squares estimators of the unknown parameters of model (8.4) minimize

$$Q(\mathbf{A}, \boldsymbol{\mu}, \boldsymbol{v}, \boldsymbol{\omega}) = \sum_{t=1}^{n} \left| y(t) - \sum_{k=1}^{p} A_k (1 + \mu_k e^{i v_k t}) e^{i \omega_k t} \right|^2, \tag{8.6}$$

with respect to $\mathbf{A} = (A_1, \ldots, A_p)$, $\boldsymbol{\mu} = (\mu_1, \ldots, \mu_p)$, $\boldsymbol{v} = (v_1, \ldots, v_p)$, $\boldsymbol{\omega} = (\omega_1, \ldots, \omega_p)$, subject to restriction (8.5).

Similarly the ALSEs of the unknown parameters are obtained by maximizing

$$I(\boldsymbol{v}, \boldsymbol{\omega}) = \sum_{k=1}^{p} \left\{ \frac{1}{n} \left| \sum_{t=1}^{n} y(t) e^{-i \omega_k t} \right|^2 + \frac{1}{n} \left| \sum_{t=1}^{n} y(t) e^{-i (\omega_k + v_k) t} \right|^2 \right\} \tag{8.7}$$

with respect to $\boldsymbol{v}$ and $\boldsymbol{\omega}$ under restriction (8.5). Write $(\widetilde{\omega}_k, \widetilde{v}_k)$ as the ALSE of (ω_k, v_k), for $k = 1, \ldots, p$. Then the corresponding ALSEs of the linear parameters of A_k and μ_k are estimated as

$$\widetilde{A}_k = \frac{1}{n} \sum_{t=1}^{n} y(t) e^{-i \widetilde{\omega}_k t}, \quad \widetilde{A}_k \widetilde{\mu}_k = \frac{1}{n} \sum_{t=1}^{n} y(t) e^{-i (\widetilde{\omega}_k + \widetilde{v}_k) t}. \tag{8.8}$$

Minimization of $Q(\mathbf{A}, \boldsymbol{\mu}, \boldsymbol{v}, \boldsymbol{\omega})$ is a $2p$-dimensional optimization problem. But, as in the case of sequential method of multiple sinusoidal model, it can be reduced to $2p$, 2-D problem.

Nandi and Kundu [12] and Nandi et al. [13] established the strong consistency of the LSEs and ALSEs and obtained their asymptotic distributions for large n under restriction (8.5) and the assumption that the sequence of the error random variables follows the assumption of stationary complex-valued linear process. It has been found that the linear parameters, that is, the real and imaginary parts of the amplitudes A_k and modulation index μ_k for $k = 1, \ldots, p$ are estimated with a rate $O_p(n^{-1/2})$, whereas frequencies are estimated with rate $O_p(n^{-3/2})$.

8.4 Fundamental Frequency Model

The fundamental frequency model is widely used in different fields of science and technology. It is a special case of general multiple sinusoidal model where different periodic components are corresponding to a particular frequency. The model takes the following form;

$$y(t) = \sum_{j=1}^{p} \rho_j^0 \cos(tj\lambda^0 - \phi_j^0) + X(t); \tag{8.9}$$

$$= \sum_{j=1}^{p} \left[A_j^0 \cos(tj\lambda^0) + B_j^0 \sin(tj\lambda^0) \right] + X(t) \quad \text{for} \quad t = 1, \ldots, n. \tag{8.10}$$

Here, $\rho_j^0 > 0$, $j = 1, \ldots, p$ are unknown amplitudes; $\phi_j^0 \in (-\pi, \pi)$, $j = 1, \ldots, p$ are unknown phases; and $\lambda^0 \in (0, \pi/p)$ is the unknown frequency; $\{X(t)\}$ is a sequence of error random variables; for $j = 1, \ldots, p$, $A_j = \rho_j^0 \cos(\phi_j^0)$ and $B_j = -\rho_j^0 \sin(\phi_j^0)$. Most of the theoretical works involving fundamental frequency model, available in the literature, are derived under Assumption 3.2. The effective frequency corresponding to j-th sinusoidal component present in $y(t)$ is $j\lambda^0$, which is of the form of harmonics of the fundamental frequency λ^0, hence the above model (8.9) has been named as *fundamental frequency model*. The model is also known as *harmonic regression signal plus noise model*.

Model (8.9) or (8.10) is a special case of the multiple sinusoidal model where $\lambda_j^0 = j\lambda^0$. This particular form of the sinusoidal model is widely used. Brown [14] discussed the asymptotic properties of a Weighted Least Squares Estimator (WLSE) defined through periodogram functions and a continuous even function under some additional assumptions. Quinn and Thomson (QT) [15] proposed an estimation method, we call it QT estimator, which is based on the weighted sum of the periodogram functions of the observed data at the harmonics. The weights are equal to the reciprocal of the spectrum of the error process at the harmonics of the fundamental frequency. This is quite a strong assumption because spectrum is usually unknown and hence a consistent estimation is required, which itself is a difficult problem. Kundu and Nandi [16] studied the theoretical properties of the LSEs and ALSEs, under Assumption 3.2, that is, the error process is a stationary linear process.

The LSEs of the unknown parameters of model (8.9) are obtained by minimizing the residual sum of squares;

$$Q(\boldsymbol{\theta}) = \sum_{t=1}^{n} \left[y(t) - \sum_{j=1}^{p} \rho_j \cos(tj\lambda - \phi_j) \right]^2, \tag{8.11}$$

with respect to the parameter vector $\boldsymbol{\theta} = (\rho_1, \ldots, \rho_p, \phi_1, \ldots, \phi_p, \lambda)$. Let $\widehat{\boldsymbol{\theta}} = (\widehat{\rho}_1, \ldots, \widehat{\rho}_p, \widehat{\phi}_1, \ldots, \widehat{\phi}_p, \widehat{\lambda})$ be the LSE of $\boldsymbol{\theta}^0 = (\rho_1^0, \ldots, \rho_p^0, \phi_1^0, \ldots, \phi_p^0, \lambda^0)$, that minimizes $Q(\boldsymbol{\theta})$ with respect to $\boldsymbol{\theta}$. We observe that λ is the only non-linear parameter and $\rho_1, \ldots \rho_p$ and $\phi_1, \ldots, \phi_p$ are either linear parameters or can be expressed in terms of the linear parameters. Hence, using separable regression technique of Richards [17], one can explicitly write the LSEs of $\rho_1^0, \ldots \rho_p^0$ and $\phi_1^0, \ldots, \phi_p^0$ as functions of λ only. Therefore, it boils down to a 1-D minimization problem.

The strong consistency of the LSE of λ^0, as well as the other parameters under Assumption 3.2, can be proved following similar techniques as the multiple sinusoidal model. The asymptotic distribution is obtained using multivariate Taylor series and first-order approximation and is stated in the following theorem.

Theorem 1 *Under Assumption 3.2,*

$$\sqrt{n}\left[(\widehat{\rho}_1 - \rho_1^0), \ldots, (\widehat{\rho}_p - \rho_p^0), (\widehat{\phi}_1 - \phi_1^0), \ldots, (\widehat{\phi}_p - \phi_p^0), n(\widehat{\lambda} - \lambda^0)\right]$$
$$\rightarrow \mathcal{N}_{2p+1}(0, 2\sigma^2 \mathbf{V})$$

as $n \rightarrow \infty$. The variance–covariance matrix $\mathbf{V}$ is as follows:

$$\mathbf{V} = \begin{bmatrix} \mathbf{C} & 0 & 0 \\ 0 & \mathbf{CD}_{\rho^0}^{-1} + \dfrac{3\delta_G \mathbf{LL}^T}{\left(\sum_{j=1}^p j^2 \rho_j^{0^2}\right)^2} & \dfrac{6\delta_G \mathbf{L}}{\left(\sum_{j=1}^p j^2 \rho_j^{0^2}\right)^2} \\ 0 & \dfrac{6\delta_G \mathbf{L}^T}{\left(\sum_{j=1}^p j^2 \rho_j^{0^2}\right)^2} & \dfrac{12\delta_G}{\left(\sum_{j=1}^p j^2 \rho_j^{0^2}\right)^2} \end{bmatrix} \tag{8.12}$$

where

$$\mathbf{D}_{\rho^0} = diag\{\rho_1^{0^2}, \ldots, \rho_p^{0^2}\}, \quad \mathbf{L} = (1, 2, \ldots, p)^T, \tag{8.13}$$

$$\delta_G = \mathbf{L}^T \mathbf{D}_{\rho^0} \mathbf{CL} = \sum_{j=1}^p j^2 \rho_j^{0^2} c(j), \quad \mathbf{C} = diag\{c(1), \ldots, c(p)\}, \tag{8.14}$$

$$c(j) = \left|\sum_{k=0}^\infty a(k) e^{-ijk\lambda^0}\right|^2. \tag{8.15}$$

In case of fundamental frequency model, the LSEs of the amplitudes only depend on the fundamental frequency λ^0 through $c(j)$. Unlike the multiple sinusoidal model, the frequency estimator asymptotically depends on all the ρ_j, $j = 1, \ldots, p$.

The ALSE of λ^0, say $\widetilde{\lambda}$, is obtained by maximizing $I_S(\lambda)$, the sum of the periodogram functions at $j\lambda$, $j = 1, \ldots, p$, defined as follows:

$$I_S(\lambda) = \frac{1}{n} \sum_{j=1}^p \left|\sum_{t=1}^n y(t) e^{itj\lambda}\right|^2. \tag{8.16}$$

The ALSEs of the other parameters are estimated as;

$$\tilde{\rho}_j = \frac{2}{n} \left| \sum_{t=1}^{n} y(t) e^{itj\tilde{\lambda}} \right|, \quad \tilde{\phi}_j = arg \left\{ \frac{1}{n} \sum_{t=1}^{n} y(t) e^{itj\tilde{\lambda}} \right\} \quad (8.17)$$

for $j = 1, \ldots, p$. Therefore, similar to LSE, estimating ALSE of λ^0 involves 1-D optimization and once $\tilde{\lambda}$ is obtained, ALSEs of the other parameters are estimated using (8.17).

Similar to the case of general sinusoidal model, ALSEs of the unknown parameters of the fundamental frequency model are strongly consistent (Kundu and Nandi [16]) and for large n have the same distribution as the LSEs.

Write model (8.9) as $y(t) = \mu(t; \theta) + X(t)$, where θ is same as defined in case of LSEs and

$$I_y(\omega_j) = \frac{1}{2\pi n} \left| \sum_{t=1}^{n} y(t) \exp(it\omega_j) \right|^2, \quad I_\mu(\omega_j, \theta) = \frac{1}{2\pi n} \left| \sum_{t=1}^{n} \mu(t; \theta) \exp(it\omega_j) \right|^2,$$

$$I_{y\mu}(\omega_j, \theta) = \frac{1}{2\pi n} \left(\sum_{t=1}^{n} y(t) \exp(it\omega_j) \right) \overline{\left(\sum_{t=1}^{n} \mu(t; \theta) \exp(it\omega_j) \right)},$$

where $\{\omega_j = 2\pi j/n; j = 0, \ldots, n-1\}$ are Fourier frequencies. Then the WLSE of θ^0, say $\widehat{\theta}$, minimizes the following objective function;

$$S_1(\theta) = \frac{1}{n} \sum_{j=0}^{n-1} \left[\left\{ I_y(\omega_j) + I_\mu(\omega_j; \theta) - 2Re(I_{y\mu}(\omega_j; \theta)) \right\} \phi(\omega_j) \right], \quad (8.18)$$

where $\phi(\omega)$ is a continuous even function of ω and it satisfies $\phi(\omega) \geq 0$ for $\omega \in [0, \pi]$.

WLSEs of the unknown parameters are strongly consistent and asymptotically normally distributed under Assumption 3.2 and some more regularity conditions. The asymptotic variance–covariance matrix depends on the chosen function $\phi(\omega)$ and the spectrum $f(\omega)$ of the error process $\{X(t)\}$, see Hannan [18], Hannan [19] and Brown [14].

The QT estimator of λ^0, proposed by Quinn and Thomson [15], say $\tilde{\tilde{\lambda}}$, minimizes

$$Q(\lambda) = \frac{1}{n} \sum_{j=1}^{p} \frac{1}{f(j\lambda)} \left| \sum_{t=1}^{n} y(t) e^{itj\lambda} \right|^2, \quad (8.19)$$

where

$$f(\lambda) = \frac{1}{2\pi} \sum_{h=-\infty}^{\infty} e^{-ih\lambda} \gamma(h) \qquad (8.20)$$

is the spectral density function or the spectrum of the error process with auto-covariance function $\gamma(.)$. Under Assumption 3.2, the spectrum of $\{X(t)\}$ has the form $f(\lambda) = \left| \sum_{j=0}^{\infty} a(j) e^{-ij\lambda} \right|^2$ and it is assumed that the spectrum of the error process is known and strictly positive on $[0, \pi]$. When the spectrum is unknown, $f(j\lambda)$ in (8.19) is replaced by its estimate. The QT estimators of the other parameters are same as the ALSEs given in (8.17). In case of QT estimator, $Q(\lambda)$ is a weighted sum of the squared amplitude estimators of model (8.9) at the harmonics $j\lambda, j = 1, \ldots, p$ and the weights are inversely proportional to the spectral density of the error random variables at these frequencies. Hence, $Q(\lambda)$ coincides with $I_S(\lambda)$ when $\{X(t)\}$ is a sequence of uncorrelated random variables. Similar to WLSE, one can term the QT estimator as Weighted ALSE.

Quinn and Thomson [15] established the strong consistency and asymptotic distribution of the QT estimator based on the assumption that the error process is ergodic and strictly stationary and the spectral density function $f(\cdot)$ at $j\lambda$ is known.

Note A comparison of the asymptotic dispersion matrix of the LSE $\widehat{\theta}$ and QT estimator $\widetilde{\theta}$ is available in Nandi and Kundu [16]. It has been noted that the asymptotic variances of the estimators of $\rho_j^0, j = 1, \ldots, p$ are same in both the cases, whereas in case of the estimators of λ^0 and $\phi_j^0, j = 1, \ldots, p$, large sample variances are different when $p > 1$. For $p > 1$, the asymptotic variances of the QT estimators are smaller than the corresponding asymptotic variances of the LSEs or ALSEs of λ^0 and ϕ_j^0 for all j, if $f(j\lambda)$ at different j are distinct. Hence, the QT estimators have certain advantages over the LSEs or ALSEs in terms of large sample variances. However, Nandi and Kundu [16] remarked that in practice QT estimators may not behave that well.

8.4.1 Test for Harmonics

The fundamental frequency model is nothing but the multiple sinusoidal model with a constraint that the frequencies are harmonics of a fundamental frequency. Once it is known that the frequencies appear at $\lambda^0, 2\lambda^0, \ldots, p\lambda^0$, the total number of parameters reduces to $2p + 1$ with only one non-linear parameter, λ^0. If p is large, this model has a substantial advantage over multiple frequency model in terms of computational difficulty. Therefore, a test of $H_0: \lambda_j = j\lambda, j = 1, \ldots, p$ against H_A: not H_0 is required where λ^0 is unknown. Quinn and Thomson [15] considered likelihood ratio statistics which is asymptotically equivalent to

$$\chi_{QT}^2 = \frac{n}{\pi} \left\{ \sum_{j=1}^{p} \frac{J(\widehat{\lambda}_j)}{f(\widehat{\lambda}_j)} - \sum_{j=1}^{p} \frac{J(j\widetilde{\widetilde{\lambda}})}{f(j\widetilde{\widetilde{\lambda}})} \right\}, \quad J(\lambda) = \left| \frac{1}{n} \sum_{t=1}^{n} y(t) e^{it\lambda} \right|^2. \quad (8.21)$$

Here $\widehat{\lambda}_j$, $j = 1, \ldots, p$ are the LSEs of λ_j^0, $j = 1, \ldots, p$ under H_A and $\widetilde{\widetilde{\lambda}}$ is the QT estimator of λ^0.

Following Quinn and Thomson [15], Nandi and Kundu [16] also used a similar statistic χ_{NK}^2, which has the same form as χ_{QT}^2, with $\widetilde{\widetilde{\lambda}}$ replaced by $\widehat{\lambda}$, the LSE of λ^0 under H_0. It has been shown that under H_0 and some regularity conditions (Quinn and Thomson [15], Nandi and Kundu [16]) χ_{QT}^2 and χ_{NK}^2 are asymptotically distributed as χ_{p-1}^2 random variable and asymptotically equivalent to

$$\chi_*^2 = \frac{n^3}{48\pi} \sum_{j=1}^{p} \frac{\widehat{\rho}_j^2 \left(\widehat{\lambda}_j - j\lambda^* \right)^2}{f(j\lambda^*)}, \quad (8.22)$$

where λ^* is either the LSE $\widehat{\lambda}$ or the QT estimator $\widetilde{\widetilde{\lambda}}$ under H_0. Furthermore, under H_0, $\chi^2 = \chi_{QT}^2 + o(1)$.

A comparative study of the two tests proposed by Quinn and Thomson [15] and Kundu and Nandi [16] will be of interest and theoretical as well as empirical comparison of these tests are important and at present an open problem.

8.5 Generalized Fundamental Frequency Model

The fundamental frequency model has been generalized by the authors (i) Irizarry [20] and (ii) Nandi and Kundu [21]. Such models can be used in case there is more than one fundamental frequency.

Irizarry [20] proposed the signal plus noise model with J periodic components; for $j = 1, \ldots, J$, $s_j(t; \boldsymbol{\beta}_j)$ is the contribution of the jth fundamental frequency and is a sum of K_j sinusoidal components of the following form:

$$y(t) = \sum_{j=1}^{J} s_j(t; \boldsymbol{\beta}_j) + X(t), \quad t = 1, \ldots, n. \quad (8.23)$$

$$s_j(t; \boldsymbol{\beta}_j) = \sum_{k=1}^{K_j} \left\{ A_{j,k} \cos(k\theta_j t) + B_{j,k} \sin(k\theta_j t) \right\}, \quad (8.24)$$

$$\boldsymbol{\beta}_j = (A_{j,1}, B_{j,i}, \ldots, A_{j,K_j}, B_{j,K_j}, \theta_j).$$

Nandi and Kundu [21] considered a similar generalization with J fundamental frequencies of the form;

$$s_j(t; \boldsymbol{\beta}_j) = \sum_{k=1}^{q_j} \rho_{jk} \cos\left\{[\lambda_j + (k-1)\omega_j]t - \phi_{jk}\right\}, \qquad (8.25)$$

$$\boldsymbol{\beta}_j = (\rho_{j_1}, \dots, \rho_{j_{q_k}}, \phi_{j_1}, \dots, \phi_{j_{q_k}}, \lambda_j, \omega_j),$$

where $\lambda_j, j = 1, \dots, J$ are fundamental frequencies and the other frequencies associated with λ_j are occurring at $\lambda_j, \lambda_j + \omega_j, \dots, \lambda_j + (q_j - 1)\omega_j$. Corresponding to j-th fundamental frequency, there are q_j bunch of effective frequencies. If $\lambda_j = \omega_j$, then frequencies effectively appear at the harmonics of λ_j. Corresponding to the frequency $\lambda_j + (k-1)\omega_j$, ρ_{jk} and ϕ_{jk}, $k = 1, \dots, q_j$, $j = 1, \dots, J$ represent the amplitude and phase components, respectively, and they are also unknown. Motivation for such a generalized model came through some real datasets.

Irizarry [20] proposed a window-based weighted least squares method and developed asymptotic variance expression of the proposed estimators. Kundu and Nandi [21] studied the theoretical properties of the LSEs of the unknown parameters. LSEs are strongly consistent and asymptotically normally distributed under the assumption of stationary linear process.

8.6 Partially Sinusoidal Frequency Model

The Partially Sinusoidal Frequency Model is proposed by Nandi and Kundu [22] with the aim of analyzing data with periodic nature superimposed with a polynomial trend component. The model in its simplest form, in the presence of a linear trend, is written as

$$y(t) = a + bt + \sum_{k=1}^{p}[A_k \cos(\omega_k t) + B_k \sin(\omega_k t)] + X(t), \quad t = 1, \dots, n+1. \quad (8.26)$$

Here $\{y(t), t = 1, \dots, n+1\}$ are the observed data and a and b, unknown real numbers, are parameters of the linear trend component. The specifications in the sinusoidal part in model (8.26) are same as the multiple sinusoidal model (8.26), described in Sect. 4.36. That is $A_k, B_k \in \mathbb{R}$ are unknown amplitudes, $\omega_k \in (0, \pi)$ are the unknown frequencies. The sequence of noise $\{X(t)\}$ satisfies Assumption 3.2. The number of sinusoidal components present is p and it is assumed to be known in advance. The initial sample size is taken as $n + 1$ instead of the usual convention as n due to some technical reason.

If b is zero, model (8.26) is nothing but model (4.36) with a constant mean term. A more general model in the class of partially sinusoidal frequency models includes a polynomial of degree q instead of the linear contribution $a + bt$.

Consider model (8.26) with $p = 1$, then

$$y(t) = a + bt + A \cos(\omega t) + B \sin(\omega t) + X(t), \quad t = 1, \dots, n+1. \quad (8.27)$$

Define $y(t + 1) - y(t) = z(t)$, say for $t = 1, \ldots n$, then

$$z(t) = y(t + 1) - y(t)$$
$$= b + A[\cos(\omega t + \omega) - \cos(\omega t)] + B[\sin(\omega t + \omega) - \sin(\omega t)] + x_d(t),$$
$$(8.28)$$

where $x_d(t) = X(t + 1) - X(t)$ is the first difference of $\{X(t)\}$ and satisfies the assumption of a stationary linear process. In matrix notation, Eq. (8.28) is written as;

$$\mathbf{Z} = b\mathbf{1} + \mathbf{X}(\omega)\boldsymbol{\eta} + \mathbf{E}, \qquad (8.29)$$

with $\mathbf{Z} = (z(1), z(2), \ldots, z(n))^T$, $\mathbf{1} = (1, 1, \ldots, 1)^T$, $\mathbf{E} = (x_d(1), \ldots, x_d(n))^T$, $\boldsymbol{\eta} = (A, B)^T$, and

$$\mathbf{X}(\omega) = \begin{bmatrix} \cos(2\omega) - \cos(\omega) & \sin(2\omega) - \sin(\omega) \\ \cos(3\omega) - \cos(2\omega) & \sin(3\omega) - \sin(2\omega) \\ \vdots & \vdots \\ \cos(\omega(n + 1)) - \cos(\omega n) & \sin(\omega(n + 1)) - \sin(\omega n) \end{bmatrix}. \qquad (8.30)$$

Then for large n, b is estimated as $\widehat{b} = (\sum_{t=1}^{n} z(t))/n$ and is a consistent estimator of b, see Nandi and Kundu [22]. Plugging $\widehat{b}$ in (8.29) and using least squares method along with separable regression technique, the frequency and linear parameters are estimated as

$$\widehat{\omega} = \arg\min_{\omega} \mathbf{Z}^{*T} \left(\mathbf{I} - \mathbf{X}(\omega) \left[\mathbf{X}(\omega)^T \mathbf{X}(\omega) \right]^{-1} \mathbf{X}(\omega)^T \right) \mathbf{Z}^*, \quad \mathbf{Z}^* = \mathbf{Z} - \widehat{b}\mathbf{1}$$

$$\widehat{\boldsymbol{\eta}} = (\widehat{A}, \widehat{B})^T = \left[\mathbf{X}(\widehat{\omega})^T \mathbf{X}(\widehat{\omega}) \right]^{-1} \mathbf{X}(\widehat{\omega})^T \mathbf{Z}^*. \qquad (8.31)$$

Testing whether b is zero is an interesting problem. which has not been addressed so far.

Nandi and Kundu [22] study the strong consistency and obtain the distribution of the estimators for large sample size. It is proved that under the condition $A^{0^2} + B^{0^2} > 0$, the estimators are consistent and for large n,

$$(\sqrt{n}(\widehat{A} - A^0), \sqrt{n}(\widehat{B} - B^0), n\sqrt{n}(\widehat{\omega} - \omega^0) \xrightarrow{d} \mathcal{N}_3(\mathbf{0}, \Sigma(\theta^0))$$

where

$$\Sigma(\theta^0) = \frac{\sigma^2 c_{par}(\omega^0)}{(1 - \cos(\omega^0)) \left(A^{0^2} + B^{0^2} \right)} \begin{bmatrix} A^{0^2} + 4B^{0^2} & -3A^0 B^0 & -6B^0 \\ -3A^0 B^0 & 4A^{0^2} + B^{0^2} & 6A^0 \\ -6B^0 & 6A^0 & 12 \end{bmatrix},$$

$$c_{par}(\omega) = \left| \sum_{k=0}^{\infty} (a(k+1) - a(k))e^{-i\omega k} \right|^2. \tag{8.32}$$

Note that $c_{par}(\omega)$ takes the same form as $c(\omega)$ presented in (4.10) defined for the differenced error process $\{x_d(t)\}$. Also,

$$\mathrm{Var}(\widehat{\boldsymbol{\theta}}^1) = \frac{c_{par}(\omega^0)}{2(1 - \cos(\omega^0))c(\omega^0)} \mathrm{Var}(\widehat{\boldsymbol{\theta}}^2),$$

where $\widehat{\boldsymbol{\theta}}^1 = (\widehat{A}^1, \widehat{B}^1, \widehat{\omega}^1)^T$ denote the LSE of $(A, B, \omega)^T$ in case of model (8.27) and $\widehat{\boldsymbol{\theta}}^2 = (\widehat{A}^2, \widehat{B}^2, \widehat{\omega}^2)^T$ is the LSE of the same model without any trend component.

In case of model (8.26), the estimation technique is the same as the one component frequency plus linear trend model (8.27). One needs to estimate the frequencies and the amplitudes using the differenced data. The coefficient b is estimated as the average of the differenced data. The corresponding design matrix $\mathbf{X}$ is of the order $n \times 2p$. Because $\mathbf{X}(\omega_j)$ and $\mathbf{X}(\omega_k)$, $k \neq j$, are orthogonal matrices, $\mathbf{X}(\omega_j)^T\mathbf{X}(\omega_k)/n = \mathbf{0}$ for large n. Therefore, the parameters corresponding to each sinusoidal component can be estimated sequentially. Nandi and Kundu [22] observe that the estimators are consistent and the parameters of the jth frequency component have similar asymptotic distribution as model (8.27) and the estimators corresponding to jth component are asymptotically independently distributed as the estimators corresponding to kth estimators, for $j \neq k$.

8.7 Burst-Type Model

The burst-type signal is proposed by Sharma and Sircar [23] to describe a segment of an ECG signal. This model is a generalized model of the multiple sinusoidal model with time-dependent amplitudes of certain form. The model exhibits occasional burst and is expressed as

$$y(t) = \sum_{j=1}^{p} A_j \exp[b_j\{1 - \cos(\alpha t + c_j)\}]\cos(\theta_j t + \phi_j) + X(t), \quad t = 1, \ldots, n,$$

$$\tag{8.33}$$

where for $j = 1, \ldots, q$, A_j is the amplitude of the carrier wave, $\cos(\theta_j t + \phi_j)$; b_j and c_j are the gain part and the phase part of the exponential modulation signal; α is the modulation angular frequency; θ_j is the carrier angular frequency and ϕ_j is the phase corresponding to the carrier angular frequency θ_j; $\{X(t)\}$ is the sequence of the additive error random variables. The number of component, p, denotes the number of carrier frequencies present. The modulation angular frequency α is assumed to be same through all components. This ensures the occasional burst at regular intervals. Nandi and Kundu [24] study the LSEs of the unknown parameters under the i.i.d.

assumption of $\{X(t)\}$ and known p. Model (8.33) can be viewed as a sinusoidal model with a time-dependent amplitude $\sum_{j=1}^{q} A_j s_j(t) \cos(\theta_j t + \phi_j) + X(t)$, where $s_j(t)$ takes the particular exponential form $\exp[b_j\{1 - \cos(\alpha t + c_j)\}]$.

Mukhopadhyay and Sircar [25] proposed a similar model to analyze an ECG signal with a different representation of parameters. Nandi and Kundu [24] show the strong consistency and obtain the asymptotic distribution as normal under the assumption of i.i.d. error and $\exp\{|b^0|\} < J$, where b^0 is the true value of b. It is observed as in case of other related models that the frequencies θ_j $j = 1, \ldots, p$ and α are estimated with a rate $O_p(n^{-3/2})$ and rest of the other parameters with rate equal to $O_p(n^{-1/2})$. When $p = 1$, the estimators of the pairs of parameters (A, b), (α, c) and (θ, ϕ) are asymptotically independent of each other, whereas the estimators of the parameters in each pair are asymptotically dependent. In case $p > 1$, the estimator of α depends on parameters of all the components.

8.8 Discussion/Remarks

Apart from the models discussed in this chapter, there are several other models which have important real-life applications and can be categorized as related models of the sinusoidal frequency model. Chirp signal model is such a model which is a natural way of extending sinusoidal model where the frequency changes linearly with time rather than being constant throughout. Chirp signal model was first proposed by Djuric and Kay [26] and is quite useful in various disciplines of science and engineering, particularly in physics, sonar, radar, and communications. Besson and his coauthors study the chirp model and different variation of this model, e.g. random or time-varying amplitudes. Nandi and Kundu [27] and Kundu and Nandi [28] develop theoretical properties of the LSEs. The generalized chirp signal is also proposed by Djuric and Kay [26], where there are p non-linear parameters corresponding to a polynomial of degree p and this polynomial acts as the arguments of sine and cosine functions. Recently Lahiri et al. [29] develop an efficient algorithm to estimate the parameters of a chirp model in the presence of stationary error.

Besson and Stoica [30] consider the problem of estimating the frequency of a single sinusoid whose amplitude is either constant or time-varying and formulate a test to detect time-varying amplitudes. Bian et al. [31] propose an efficient algorithm for estimating the frequencies of the superimposed exponential signal in zero mean multiplicative as well as additive noise.

Bloomfield [32] considered the fundamental frequency model when the fundamental frequency is a Fourier frequency. Baldwin and Thomson [33] and Quinn and Thomson [15] used model (8.9) to explain the visual observation of S. Carinae, a variable star in the Southern Hemisphere sky. Greehouse et al. [34] proposed higher-order harmonics of one or more fundamental frequencies with stationary ARMA processes for the errors to study of biological rhythm data, illustrated by human core body temperature data. The harmonic regression plus correlated noise model has also been used in assessing static properties of human circadian system, see Brown and Czeisler [35] and Brown and Liuthardt [36] and examples therein. Musical sound segments produced by certain musical instruments are mathematically explained using such models. Some short duration speech data are analyzed using model (8.9), see Nandi and Kundu [16] and Kundu and Nandi [27]. Several interesting open problems exist in these areas. Further attention is needed along these directions.

References

1. Kay, S. M. (1988). *Modern spectral estimation: Theory and application*. New York: Prentice Hall.
2. Seber, A., & Wild, B. (1989). *Nonlinear Regression*. New York: Wiley.
3. Bates, D. M., & Watts, D. G. (1988). *Nonlinear regression and its applications*. New York: Wiley.
4. Kumaresan, R. (1982). *Estimating the parameters of exponential signals*. Ph.D. thesis, U. Rhode Island.
5. Kumaresan, R. & Tufts, D. W. (1982). Estimating the parameters of exponentially damped sinusoids and pole-zero modelling in noise. *IEEE Transactions on ASSP, 30*, 833–840.
6. Rao, C. R. (1988). Some results in signal detection. In S. S. Gupta & J. O. Berger (Eds.), *Decision theory and related topics, IV* (Vol. 2, pp. 319–332). New York: Springer.
7. Wu, C. F. J. (1981). Asymptotic theory of non linear least squares estimation. *Annual Statistics, 9*, 501–513.
8. Kundu, D. (1994). A modified Prony algorithm for sum of damped or undamped exponential signals. *Sankhya, 56*, 524–544.
9. Kundu, D. (1991). Assymptotic properties of the complex valued non-linear regression model. *Communications in Statistics: Theory and Methods, 20*, 3793–3803.
10. Jennrich, R. I. (1969). Asymptotic properties of the non linear least squares estimators. *Annals of Mathematical Statistics, 40*, 633–643.
11. Sircar, P., & Syali, M. S. (1996). Complex AM signal model for non-stationary signals. *Signal Processing, 53*, 35–45.
12. Nandi, S., & Kundu, D. (2002). Complex AM model for non stationary speech signals. *Calcutta Statistics Associated Bulletin, 52*, 349–370.
13. Nandi, S., Kundu, D., & Iyer, S. K. (2004). Amplitude modulated model for analyzing non stationary speech signals. *Statistics, 38*, 439–456.
14. Brown, E. N. (1990). A note on the asymptotic distribution of the parameter estimates for the harmonic regression model. *Biometrika, 77*, 653–656.
15. Quinn, B. G., & Thomson, P. J. (1991). Estimating the frequency of a periodic function. *Biometrika, 78*, 65–74.
16. Nandi, S., & Kundu, D. (2003). Estimating the fundamental frequency of a periodic function. *Statistical Methods and Applications, 12*, 341–360.
17. Richards, F. S. G. (1961). A method of maximum likelihood estimation. *Journal of the Royal Statistical. Society B, 23*, 469–475.
18. Hannan, E. J. (1971). Non-linear time series regression. *Journal of Applied Probability, 8*, 767–780.
19. Hannan, E. J. (1973). The estimation of frequency. *Journal of Applied Probability, 10*, 510–519.
20. Irizarry, R. A. (2000). Asymptotic distribution of estimates for a time-varying parameter in a harmonic model with multiple fundamentals. *Statistica Sinica, 10*, 1041–1067.
21. Nandi, S., & Kundu, D. (2006). Analyzing non-stationary signals using a cluster type model. *Journal of Statistics Planning and Inference, 136*, 3871–3903.

22. Nandi, S., & Kundu, D. (2011). Estimation of parameters of partially sinusoidal frequency model. *Statistics,*. doi:10.1080/02331888.2011.577894.
23. Sharma, R. K., & Sircar, P. (2001). Parametric modelling of burst-type signals. *Journal of the Franklin Institute, 338,* 817–832.
24. Nandi, S., & Kundu, D. (2010). Estimating the parameters of Burst-type signals. *Statistica Sinica, 20,* 733–746.
25. Mukhopadhyay, S., & Sircar, P. (1996). Parametric modelling of ECG signal. *Medical and Biological Engineering and computing, 34,* 171–174.
26. Djuric, P. M., & Kay, S. M. (1990). Parameter estimation of chirp signals. *IEEE Transactions on ASSP, 38,* 2118–2126.
27. Kundu, D., & Nandi, S. (2004). A Note on estimating the frequency of a periodic function. *Signal Processing, 84,* 653–661.
28. Kundu, D., & Nandi, S. (2008). Parameter estimation of chirp signals in presence of stationary noise. *Statistica Sinica, 18,* 187–201.
29. Lahiri, A., Kundu, D., & Mitra, A. (2012). Efficient algorithm for estimating the parameters of chirp signal. *Journal of Multivariate Analysis, 108,* 15–27.
30. Besson, O., & Stoica, P. (1999). Nonlinear least-squares approach to frequency estimation and detection for sinusoidal signals with arbitrary envelope. *Digital Signal Processing, 9,* 45–56.
31. Bian, J. W., Li, H., & Peng, H. (2009). An efficient and fast algorithm for estimating the frequencies of superimposed exponential signals in zero-mean multiplicative and additive noise. *Jounal of Statistics Computation and Simulation, 79,* 1407–1423.
32. Bloomfiled, P. (1976). *Fourier analysis of time series.* Wiley, New York: An introduction.
33. Baldwin, A. J., & Thomson, P. J. (1978). Periodogram analysis of S. Carinae. *Royal Astronomical Society of New Zealand (Variable Star Section), 6,* 31–38.
34. Greenhouse, J. B., Kass, R. E., & Tsay, R. S. (1987). Fitting nonlinear models with ARMA errors to biological rhythm data. *Statistics in Medicine, 6,* 167–183.
35. Brown, E. N., & Czeisler, C. A. (1992). The statistical analysis of circadian phase and amplitude in constant-routine core-temperature data. *Journal of Biological Rhythms, 7,* 177–202.
36. Brown, E. N., & Liuthardt, H. (1999). Statistical model building and model criticism for human circadian data. *Journal of Biological Rhythms, 14,* 609–616.

Index

D. Kundu and S. Nandi, *Statistical Signal Processing*, SpringerBriefs in Statistics   129
DOI: 10.1007/978-81-322-0628-6, © The Author(s) 2012